Modern Meteorology and Climatology

Modern Meteorology and Climatology: an introduction

T. J. Chandler M.SC., PH.D.

Professor of Geography
University College London

Nelson

Thomas Nelson and Sons Ltd
36 Park Street London W1Y 4DE
PO Box 18123 Nairobi Kenya
Thomas Nelson (Australia) Ltd
597 Little Collins Street Melbourne 3000
Thomas Nelson and Sons (Canada) Ltd
81 Curlew Drive Don Mills Ontario
Thomas Nelson (Nigeria) Ltd
PO Box 336 Apapa Lagos
Thomas Nelson and Sons (South Africa) (Proprietary) Ltd
51 Commissioner Street Johannesburg

First published in 1972

ISBN 0 17 444012 X

Printed in Great Britain by A. Wheaton & Co., Exeter

Contents

Acknowledgements

The sources of information used in the maps have been acknowledged in the figure captions. For permission to use copyright material in the plates, our thanks are due to the following:

The Mansell Collection: Plate 1
The Science Museum: Plates 2, 3, 4, 5
Camera Press Ltd and NASA: Plate 6
The Meteorological Office: Plates 7, 8, 10, 16
Associated Press Ltd.: Plate 9
Mr. G. Nicholson: Plates 11, 12, 13
McGraw-Hill Ltd.: Plate 15 (from *Snow Crystals* by W. A. Bentley and W. J. Humphreys, 1931)
Central Press Photos Ltd: Plate 17
Fox Photos Ltd: Plate 18
The Imperial War Museum and Camera Press Ltd: Plate 20
Mr. W. Harper and *Weather:* Plate 14
Aldus Books Ltd. and the Division of Radiophysics, Commonwealth Scientific and Industrial Research Organization: Plate 19

Diagrams © T. J. Chandler; cover design by Roger Bristow; book design by Jill Leman.

Chapter 1

Weather lore to weather laws

Man soon learnt the importance of the weather in controlling where and how he lived, what clothes he wore, the design of his shelters and buildings and which crops he could grow. From the beginning, an elementary knowledge of the weather was clearly necessary for his survival.

Most ancient peoples believed that the weather was governed by the caprice of the gods, although Aristotle was an early pioneer of true scientific understanding and in his book *Meteorologica*, published in the fourth century B.C., he gave fairly accurate descriptions and explanations of several meteorological phenomena. But before the seventeenth century lack of instruments meant there was little advance in true understanding, in place of which grew up an assembly of astrological and religious theories and local traditions. Many of these traditions showed careful observation of local weather which was often expressed in the form of rhymes and jingles and represented the accumulated knowledge of generations of mainly country people. Modern town dwellers are much less dependent, directly at least, upon the vagaries of the local weather since they spend much of their lives in the artificial climates of buildings and vehicles. Even so, the shape of the ground, its vegetation cover, the use man makes of the land, the design of buildings and even the way men behave are undeniably related to weather and climate. Climate can be regarded as the integral of weather, which itself includes the enormously important extreme occurrences. Climate is not, therefore, average weather.

Today we have many ingenious ways of measuring the weather throughout the depth of the atmosphere, together with an internationally-organised system of observations and rapid communications. With advances in modern theory and the enormously rapid and complex calculations made possible by electronic computers, these methodical observations have, in ths last decade or two, led to tremendous advances in our understanding of the weather and in the accuracy of our forecasts. Many old ideas have had to be abandoned as new discoveries have been made and we live in a tremendously exciting time in the history of meteorology.

Chapter 2

Measuring the atmosphere

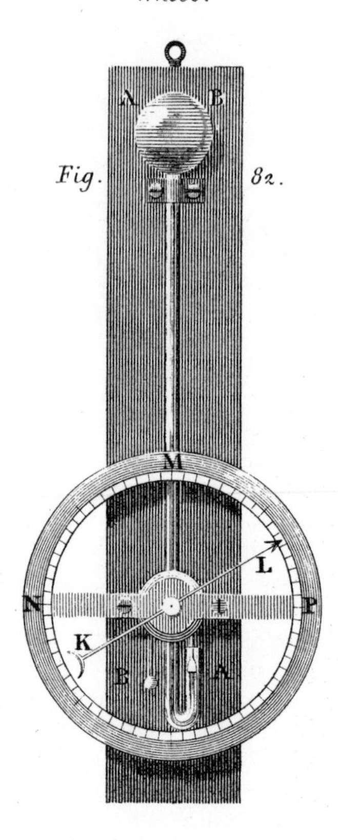

1. Robert Hooke's wheel barometer. Hooke (1635–1703) was a leading seventeenth century scientist whose major work was concerned with the application of optics, and of the theory of gravity. Within the central circle in this contemporary print, an iron ball (at point A) rests on the surface of the mercury, with a string going over the pulley to a smaller ball B. As the mercury rises in the tube so the pointer L moves towards point M. As the mercury level drops, the pointer moves towards P.

Meteorology as a science was born in 1643 with the invention of the barometer by the Italian, Evangelista Torricelli. He used it to demonstrate the existence of atmospheric pressure, but soon afterwards, fluctuations of pressure were noticed and these were seen to bear upon the weather. Unfortunately, perhaps, the English scientist Robert Hooke suggested a far closer relationship between pressure and weather than actually exists when in 1670 he made the first 'weather glass' (Plate 1) on which, as on all the millions of subsequent copies, he equated low pressure with stormy conditions and rain, and high pressure with dry, fair conditions.

The first liquid-in-glass thermometer was invented at about the same time as the barometer, probably by Galileo, and the first hygrometer, based upon the expansion and contraction of human hair with changes in humidity, followed soon afterwards.

Three basic elements of the atmosphere—pressure, temperature and humidity—could now be measured, but although correlations between these and other aspects of the weather such as winds, rain and cloud began to be made locally, the number of observations was so small, the techniques of observation so varied and the rapid assembly of the data so rarely possible that few spatial interpretations were possible and the patterns of weather and climate, so basic to our present understanding of the atmosphere, were hardly known.

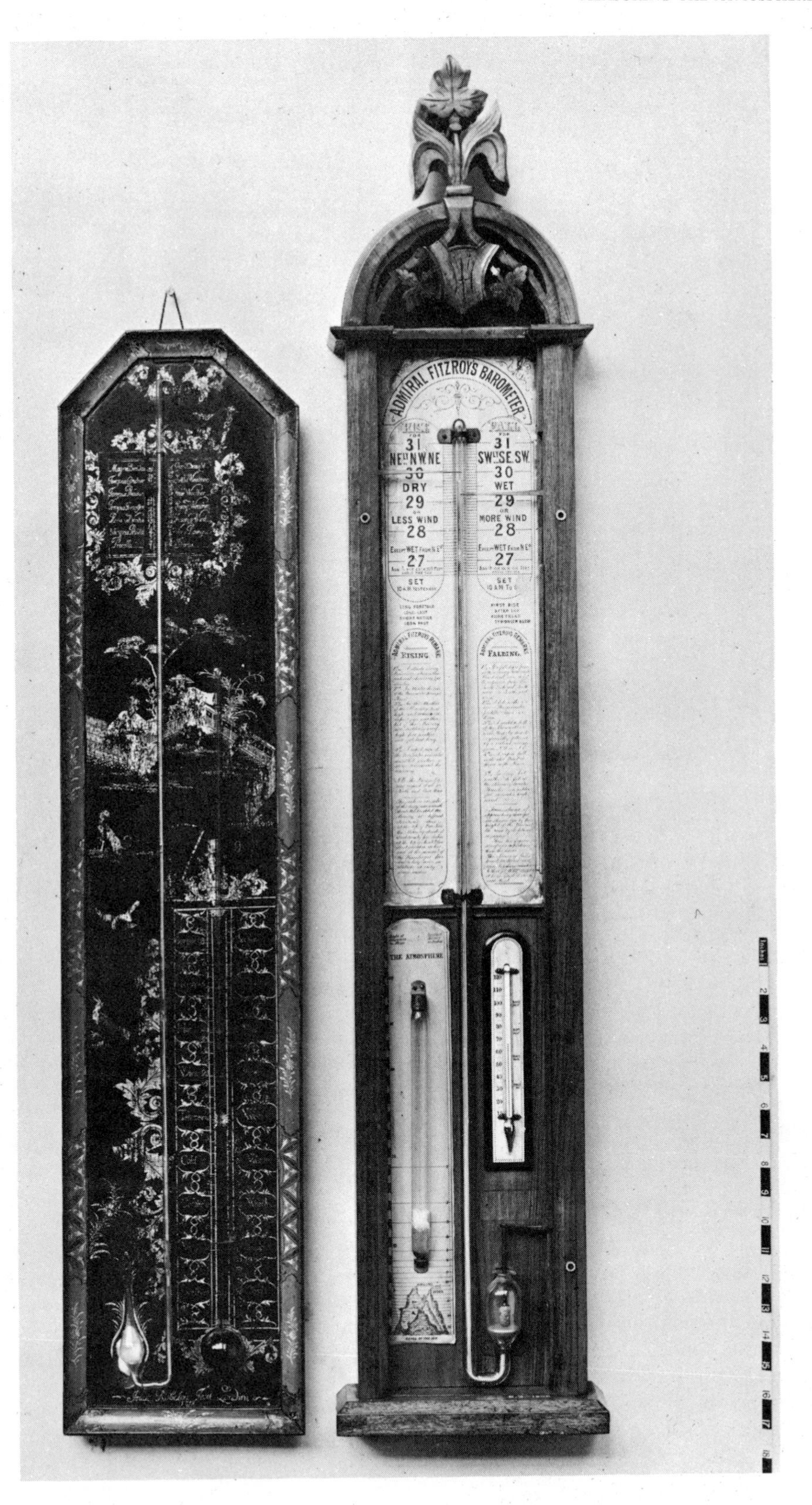

2. Robert Fitzroy's barometer. Admiral Fitzroy (1805–1865) was commander of HMS *Beagle* on the 1831–6 survey of the South American coast, carrying Charles Darwin as an observer. Fitzroy became chief of the new meteorological department of the Board of Trade on its foundation in 1864.

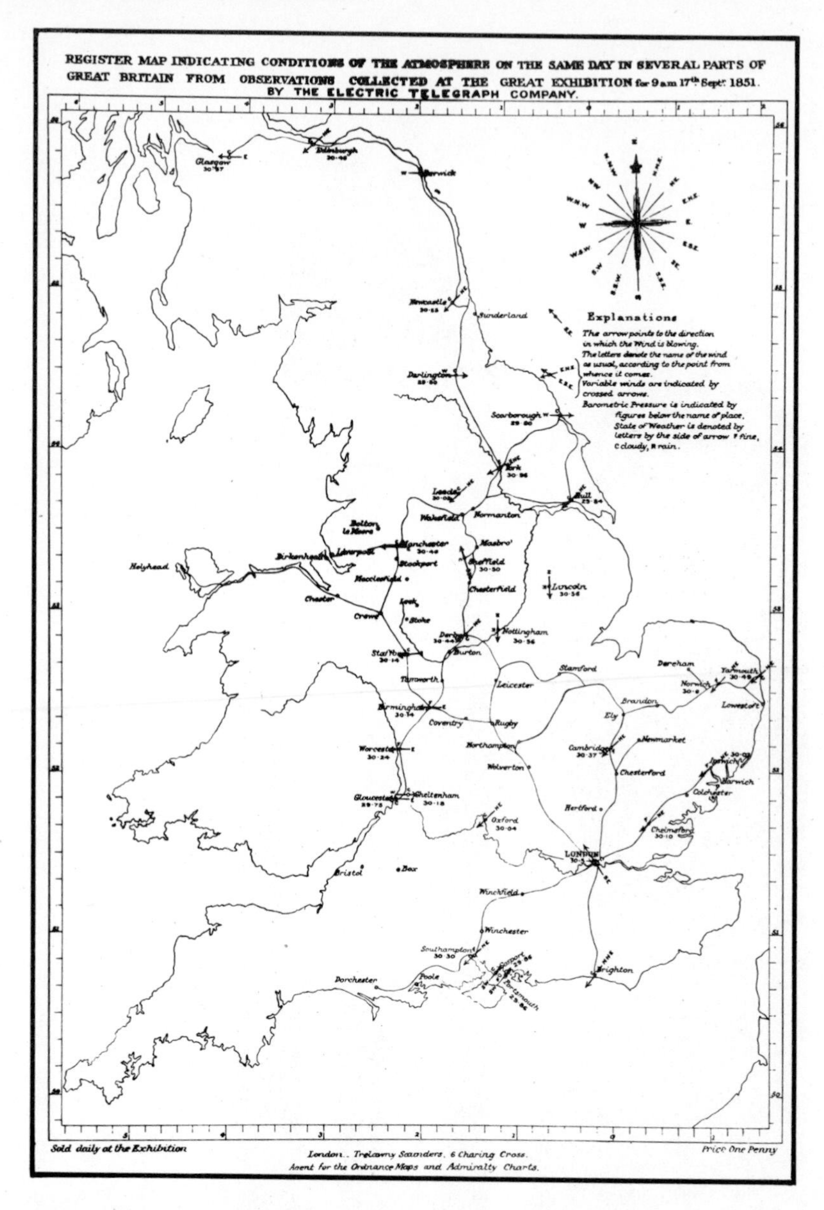

3. Weather charts compiled from observations taken in different parts of the country became available soon after the development of the national electric telegraph service. This example, for 17 September, 1851, was sold at the great Exhibition, and was compiled by one of the several competing telegraph companies.

The first weather map of which we have any record was the one made in 1686 by the English astronomer Edmond Halley. It showed the average wind directions between 30°N. and S. and was therefore, more strictly a climate map. The first true synoptic weather chart showing simultaneous conditions over a fairly large area had to wait upon the invention and development of the telegraph in the 1840s (Plates 3–5). This enabled the rapid assembly of the information necessary to build up a series of instantaneous pictures of the weather. As a result of this it was soon noticed how areas of similar weather moved across the charts, enlarging and diminishing, intensifying and weakening. At the same time many more surface stations were set up to record the weather using internationally agreed methods, although even today there are still large areas of the world, more particularly in the southern hemisphere, where observations are few and far between. Values for ocean areas are particularly scanty in spite of readings by anchored weather ships and

Septr. 3rd

WEATHER REPORT.

1860.

At 9 A.M.

	B.	E.	M.	D.	F.	C.	I.
Aberdeen							
Greenock	30·07	55	52	WSW	2	1	b
Berwick							
Copenhagen							
Portrush							
G[illegible]							
Hu[illegible]	30·06	54	52	W	2	6	o
Liverpool							
Queenstown							
Helder							
Yarmouth	30·06	63	59	NW	2	5	c
London	30·13	58	54	W	2	2	b
~~Nore~~ Dunkirk	30·15	59	52	WSW	0	1	b
Dover							
Portsmouth	29·96	59	58	SW	3	3	bc
Plymouth	30·06	60	—	NNW	2	8	oc
Cherbourg	30·11	61	55	WNW	0	1	b
~~Penzance~~ Havre		57	—	—	—	2	bc
Jersey	30·15	59	56	NNW	2	2	bc
Brest	30·07	52	—	NW	0	9	oc
Bayonne							
Lisbon							

EXPLANATION.

B.—Barometer corrected and reduced to 32° at sea-level (mean). E.—Exposed (but shaded) thermometer. M.—Moistened bulb (for evaporation and dew point). D.—Direction of wind (true). F.—Force (0 to 12). C.—Cloud (1 to 9) proportion. I.—Initial letters: b.—blue sky; c.—clouds (detached); f.—fog; h.—hail; l.—lightning; m.—misty (hazy); o.—overcast (dull); r.—rain; s.—snow; t.—thunder.

NOTE.—A letter repeated augments—thus, r r much rain.

4. The electric telegraph enabled weather reports, like this for 3 September, 1860, to be compiled for a variety of areas . . .

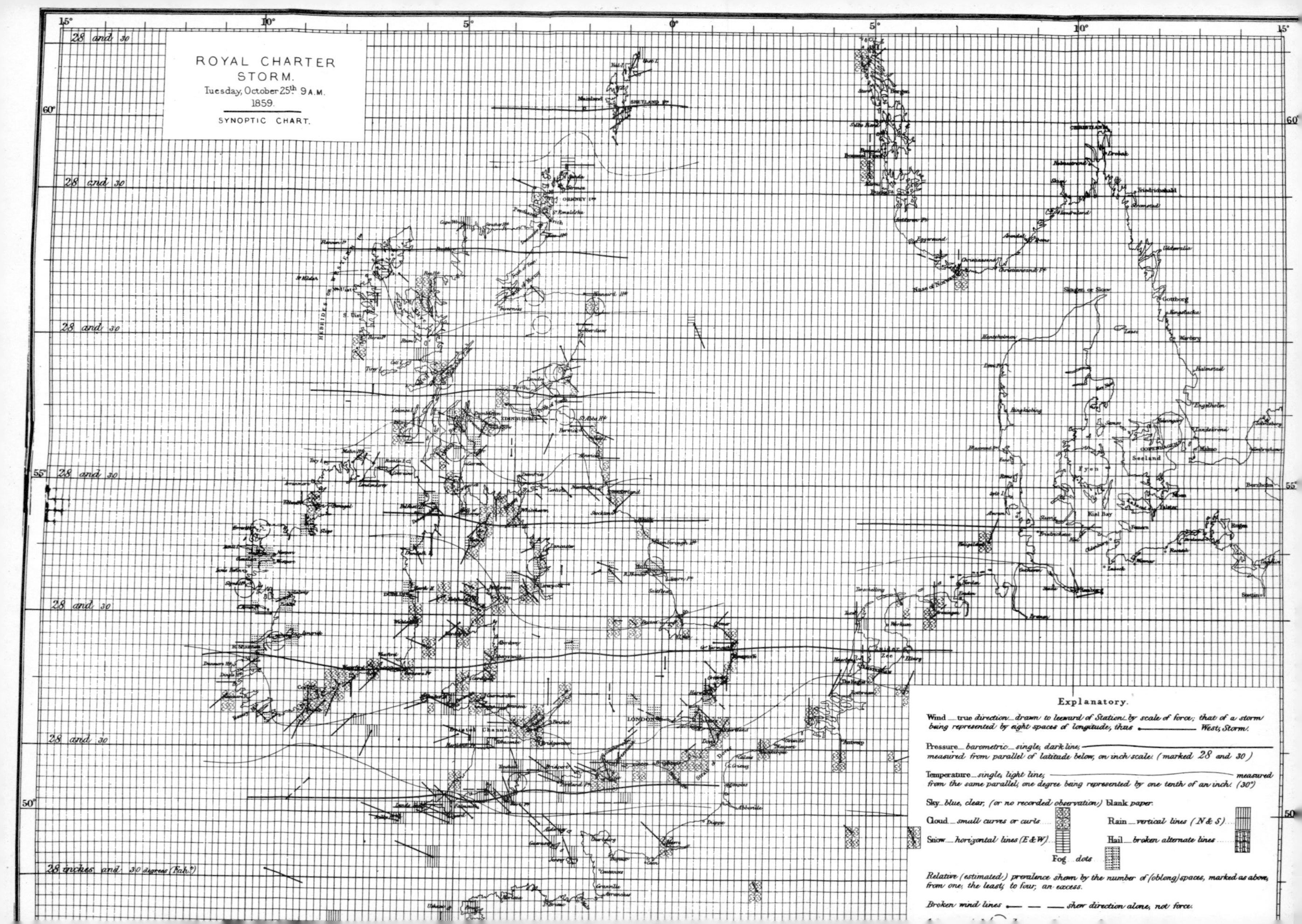
ROYAL CHARTER
STORM.
Tuesday, October 25th 9 A.M.
1859.
SYNOPTIC CHART.
Explanatory.
Wind _ true direction _ drawn to leeward of Station, by scale of force; that of a storm being represented by eight spaces of longitude, thus ——— West, Storm.
Pressure _ barometric _ single, dark line; measured from parallel of latitude below, on inch scale: (marked 28 and 30)
Temperature _ single, light line; measured from the same parallel; one degree being represented by one tenth of an inch: (30°)
Sky _ blue, clear, (or no recorded observation) blank paper.
Cloud _ small curves or curls
Rain _ vertical lines (N & S)
Snow _ horizontal lines (E & W)
Hail _ broken alternate lines
Fog _ dots
Relative (estimated) prevalence shewn by the number of (oblong) spaces, marked as above, from one, the least, to four, an excess.
Broken wind lines — — — shew direction alone, not force.
28 and 30
28 inches and 30 degrees (Fah.t)
LONDON
EDINBURGH
DUBLIN
COPENHAGEN
CHRISTIANIA
SHETLAND Is.
ORKNEY Is.
HEBRIDES OR WESTERN Is.
Bristol Channel
Seeland
Fyen
Kiel Bay
Zuider Zee

hundreds of merchant ships, and this is why the World Meteorological Organization is currently engaged to increase our coverage of these areas and further improve the assembly and rapid international exchange of information.

Weather satellites (Plate 6) have helped enormously, for they record not only cloud patterns (Plates 7–8) but also ground temperature and the atmosphere's vertical temperature profile. Instrument-carrying constant level balloons, circling at various heights above the southern oceans, are another recently-introduced technique, greatly adding to our knowledge of the atmosphere.

Today it is possible to explore the atmosphere in far greater detail than ever before, providing information by which to evolve and check theories. These theories, often expressed in mathematical terms, can then be used to forecast the weather, perhaps, in a general way, over several months though detailed forecasts are unlikely to be possible for more than a few days ahead, at least for some time.

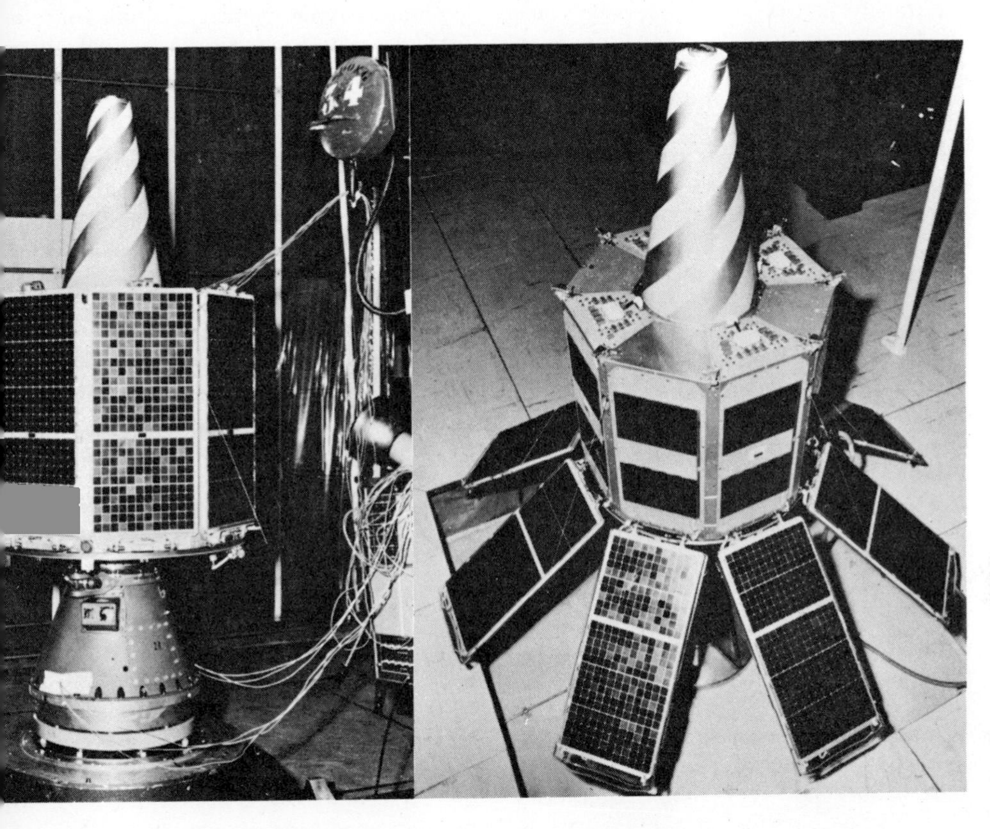

6. Modern weather satellite. This one, the French *Eole* satellite launched in August, 1971, weighs 84 kilogrammes and collects information on winds, temperatures and pressures from instrumented balloons flying at an altitude of 11,850 metres, in the southern hemisphere

5. . . . and so synoptic weather charts *(opposite page)*, showing the weather conditions across wider areas, as in this chart for 25 October, 1859, could be assembled.

Chapter 3

Structure of the atmosphere

The absolute depth of the ocean of air above the earth has been somewhat arbitrarily placed at about 600 miles (1000 kilometres). It is impossible to be precise about this because the atmosphere thins progressively without any sharp boundary, although at great heights there are no more than slight traces of atmospheric gases. Half of the atmosphere's mass lies in the 3½ miles (5·5 km) nearest the earth and more than 99 per cent within 25 miles (40 km). At 60 miles (100 km) up there is a near vacuum with only one millionth of the pressure at ground level. The lowest atmosphere is thus very compressed with a sharp vertical pressure gradient and the outward force produced by this gradient is, in the absence of vertical motion, balanced by the inward force of gravity. If the balance is disturbed by changes of pressure, then there is uplift or descent of air.

In contrast to the complicated physics of the atmosphere, its chemistry is comparatively simple. It is made up of two types of gases: primarily, it contains the so-called permanent gases dominated by nitrogen (78 per cent) and oxygen (21 per cent), plus many others such as hydrogen and helium in very small quantities. Secondly, there are the variable gases whose concentrations vary in both space and time. Two of the most important of these are ozone and water vapour. Ozone is found mainly between heights of 20 and 50 miles (30 and 80 km) and although even here the concentrations are very small, the gas is vital to atmospheric processes and life on earth since it absorbs harmful short-wave radiation from the sun. Water vapour on the other hand is almost entirely concentrated in the first 6 to 9 miles (10 to 15 km) of the atmosphere and becomes visible only when condensed. If it were all condensed and precipitated, it would amount to about an inch (25·5 mm) of rain over the entire earth or about 900 tons of water per acre (2250 tons per hectare).

Certain of the variable gases, sulphur dioxide and carbon dioxide for instance, seem to undergo periodic changes in concentration. Both gases are partly by-products of the burning of coal and oil, and although small in amount, changes in their concentrations affect the heat exchanges which control the temperature of the atmosphere.

The basic thermal stratification of the lower atmosphere was first demonstrated at the end of the nineteenth century by the French meteorologist Teisserenc de Bort, using balloons to carry recording instruments to about 9 miles (15 km) above Paris. He showed that temperatures generally decreased up to heights of about 7 miles (11 km) and then increased with height. This break of trend was later shown to be a universal feature of the lower atmosphere though varying in height from about 5 miles (8 km) above polar regions to 10 miles (16 km) above the Equator. It is known as the **tropopause** and more recent research has shown there to be two and sometimes three distinct tropopause levels separated by sudden changes in height (Fig. 1). The tropopause divides the troposphere near the earth from the next highest layer above, the **stratosphere.** In the former, temperatures generally decrease with height, in the latter they increase upwards (Fig. 1). Though most weather processes of consequence to the earth occur in the troposphere, the lower stratosphere (and perhaps even higher levels) now appears to play some part in these processes. In any case most

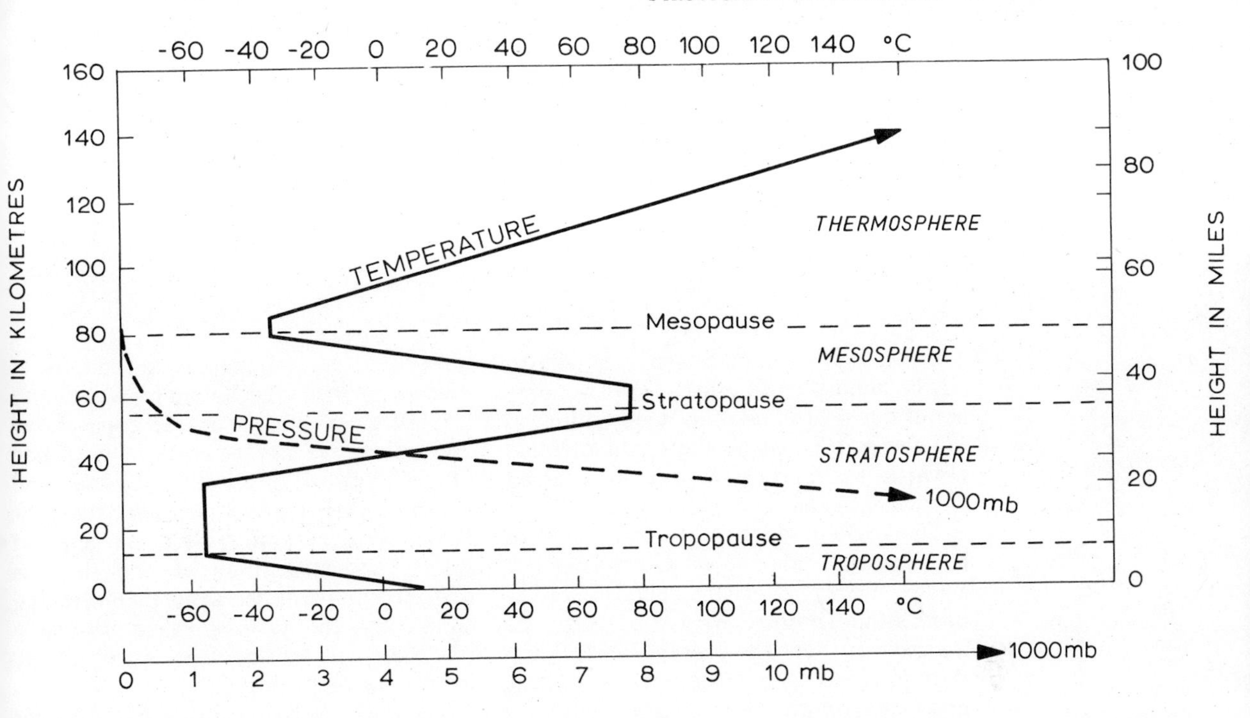

modern civil jets fly at heights of 35 000 to 40 000 ft (10 000 to 12 000 m), that is, in the lower stratosphere. The new supersonic jets will fly even higher, so that stratopheric winds and weather are now of great practical importance.

Near the top of the stratosphere where ozone strongly absorbes short-wave solar radiation, temperatures are roughly the same as those near the earth. At about 35 miles (56 km) up, however, a thermal maximum occurs, above which temperatures fall rapidly with height. The thermal maximum is known as the **stratopause** and separates the stratosphere below from the **mesosphere** above. Above the mesosphere in which temperatures fall with height is the **thermosphere** or **ionosphere** in which they again rise, this time to temperatures in excess of 1000°C. Little is sknown about conditions at these levels above 50 miles (80 km) but satellites and rockets are already exploring them and enabling us to build up a truer picture of the earth's 'ocean of air'.

Fig. 1 The variation of temperature and pressure with height. Changes in the rise or fall of temperature define the several layers of the atmosphere (troposphere, etc.). After diagram in Petterssen, S., *Introduction to Meteorology* (McGraw-Hill, 1958).

Chapter 4

The atmospheric heat engine

The sun is the ultimate source of almost all the atmosphere's energy; even so, only about one part in two thousand million of the energy emitted by the sun is directed towards the earth; the rest is beamed elsewhere in space.

The sun, a gaseous sphere with a surface temperature of about 6000°C., transmits its energy mainly as electro-magnetic radiation, although there is also a stream of ionized particles. The energy travels freely through space but is converted to heat upon absorption; the absorbing body is then warmed and re-radiates energy, although in different amounts and wave-lengths. Because of its lower temperatures, the earth emits much less energy and in longer wave-lengths than the sun. Thus whilst the earth's radiation lies between 4 and 50 microns (a micron is a millionth part of a metre), the sun emits 99 per cent of its energy in wave-lengths less than 4 microns. About half the sun's radiation is, in fact, in wave-lengths visible to the human eye as light, that is, between 0.38 (violet light) and 0.77 (red light) microns. Much of the remainder is in only slightly shorter (ultra-violet) or slightly longer (infra-red) wave-lengths.

The amount of solar energy received per unit area and time on a surface at right-angles to the sun's beam at the fringes of the earth's atmosphere is known as the **solar constant.** It is almost 8·38 joules (2 calories) per cm^2 per minute. The daily amount depends upon the elevation of the sun and the length of day so that the greatest amount received occurs over each pole at midsummer, more particularly over the southern pole in December, because the earth is then slightly nearer the sun (Fig. 2). The colossal input from the sun to the earth-atmosphere system is equivalent to the energy produced by 180 million large power stations (and there is no room for so many on earth). But only a little over half the energy impinging on the edge of the atmosphere reaches the earth. Almost one fifth is absorbed by the tiny particles and various gases of the atmosphere, particularly ozone, carbon dioxide, oxygen and water vapour; and about one third is scattered and reflected from air molecules, clouds and the earth so that eventually less than half the incoming solar energy warms the earth (Fig. 3).

Terrestial radiation, being much longer in wave-length than solar radiation, cannot pass so easily through the atmosphere; indeed, much is absorbed and re-radiated down again. Only a small fraction escapes directly to space, mainly through the so-called 'window' in the water vapour absorption spectrum, that is, through a narrow waveband in which there is no absorption by the water vapour in the atmosphere. In longer and shorter wave-bands there is partial or total absorption and the atmosphere thus acts like a blanket to limit severely the loss of radiant heat from the earth. Heat is passed from the earth to the atmosphere mainly by eddy (rather than molecular) conduction in the turbulent overturnings of the lower atmosphere, and by latent heat transfer, that is, evaporation and the absorption of latent heat at the earth's surface followed by condensation (in clouds) and the release of this heat into the atmosphere. Evaporation and condensation are, of course, often separated by thousands of miles. Latent-heat transfer is on average more than twice as important as eddy conduction in warming the atmosphere from the earth and, in summer in Great Britain, about half the sun's energy reaching the earth is used in evaporation.

Finally, of course, heat is radiated from

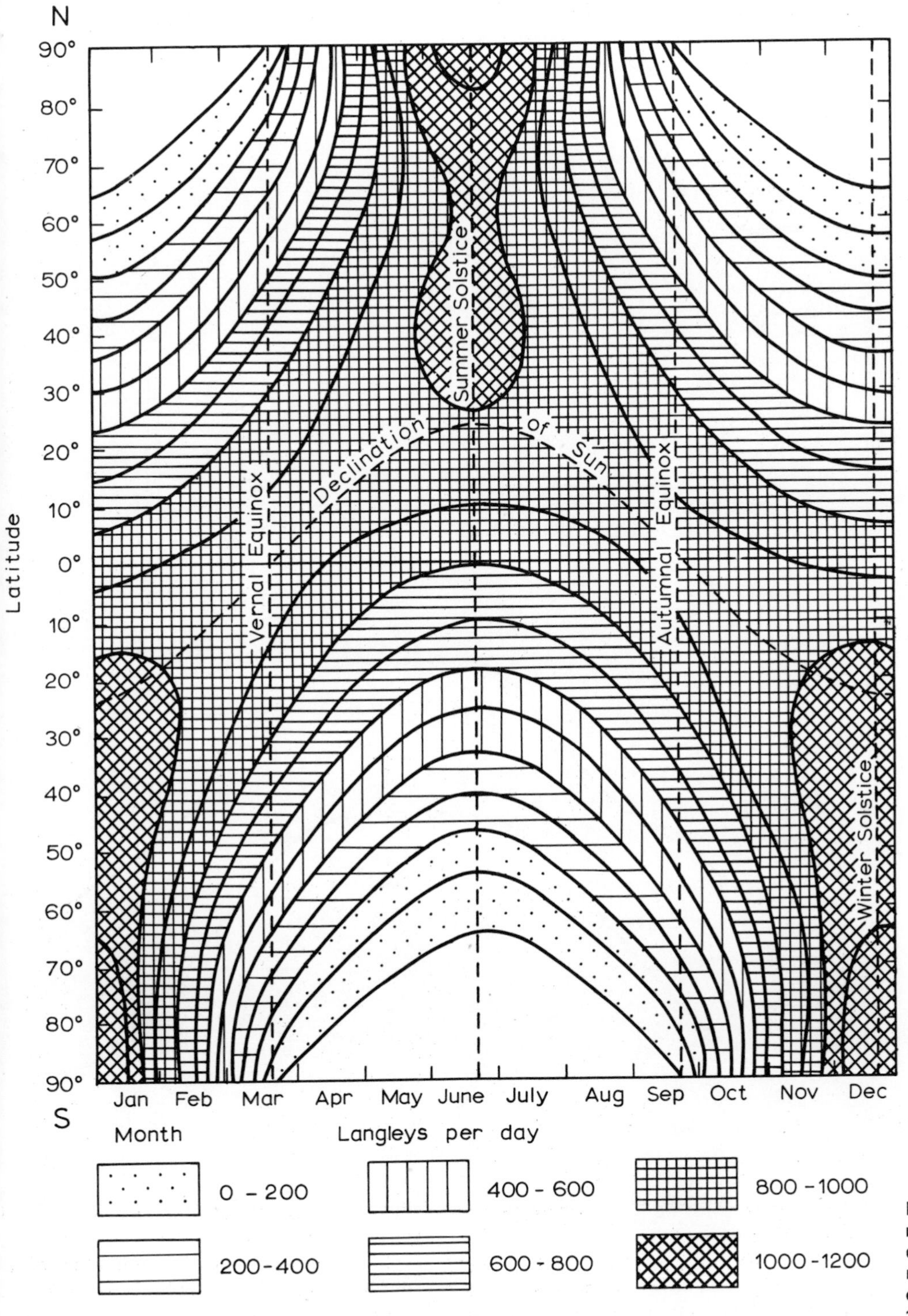

Fig. 2 Radiation received per day in different latitudes and months at the outer edge of the atmosphere. 1 Langley = 1 cal/cm^2.

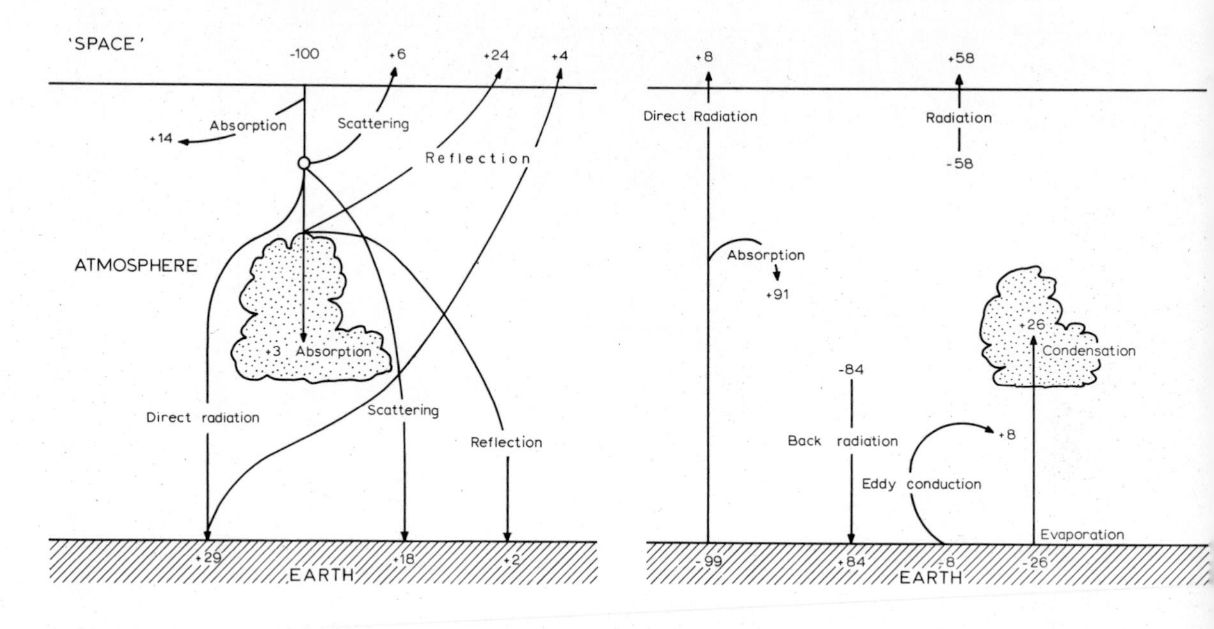

Fig. 3 The annual heat transfers between the sun and space, the atmosphere and the earth. The units are non-dimensional.

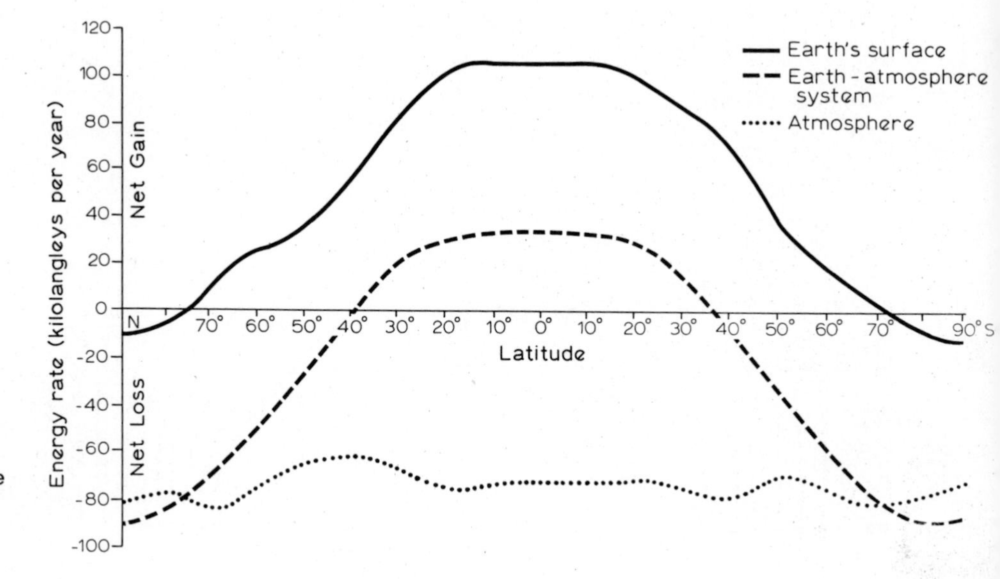

Fig. 4 Latitudinal variations in the average net radiation gain or loss by the earth's surface and by the atmosphere, separately and together. After diagram in W. D. Sellers, *Physical Climatology* (University of Chicago Press, 1965).

the outer parts of the atmosphere to space (Fig. 3).

Up to this point we have been speaking about the earth–atmosphere system as a whole and because there are only small year-to-year changes in mean temperature, we can be certain that radiation received from the sun is balanced by that lost from the earth and its atmosphere; input and output cancel each other out. But this balance does not exist for individual parts of the system. The earth's surface has a surplus of energy from radiation everywhere except in polar areas (Fig. 4) where there is an enormous reflection and loss of solar energy from clouds and surface ice

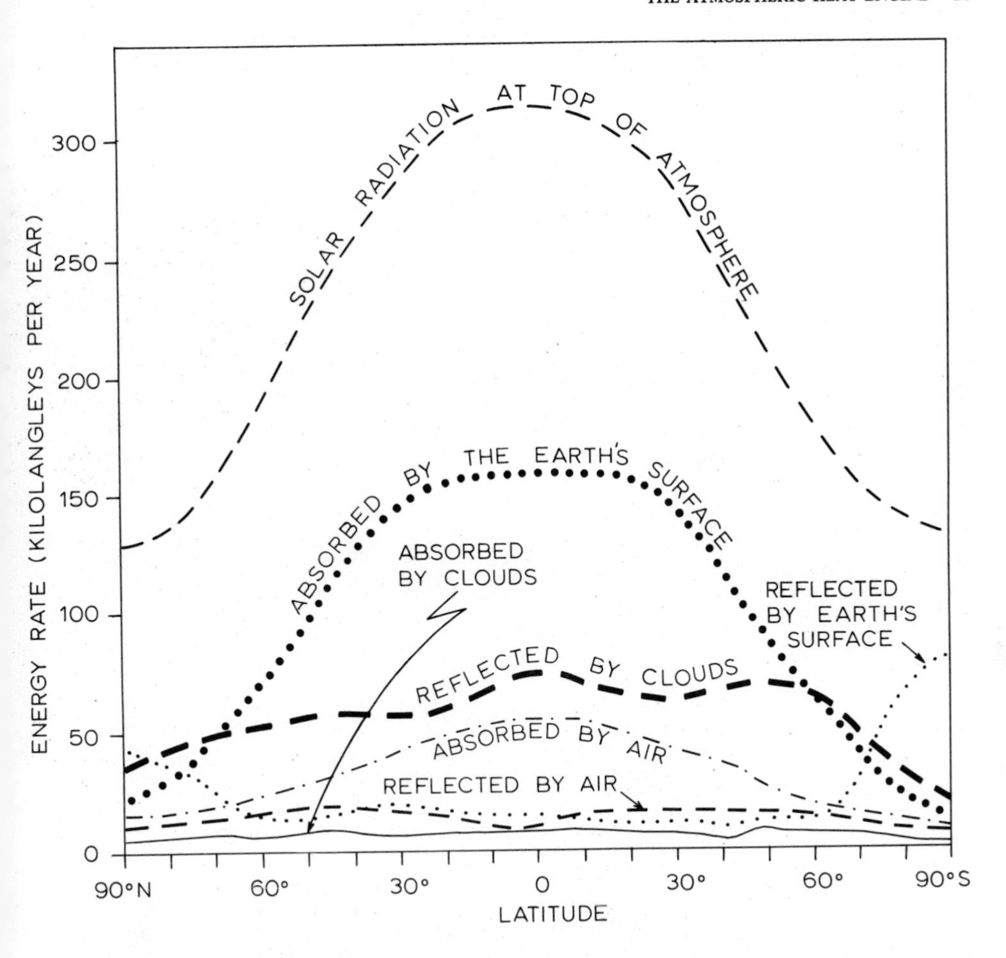

Fig. 5 Latitudinal variations in the average annual disposition of solar radiation. After diagram in W. D. Sellers, *Physical Climatology* (University of Chicago Press, 1965).

(Fig. 5). The atmosphere, however, everywhere loses more radiant energy than it gains. Thus in the absence of compensating exchanges, the earth would be constantly warm and the atmosphere cool. If we now consider the amount of net surplus and deficit in different latitudes, and combining the earth and overlying air, we find that equatorwards of about 38° latitude there is a surplus and polewards, a deficit (Fig. 4). In comparison, measured radiation losses to space from the outer edges of the atmosphere are almost constant in the different latitudes. And so we have the fundamental situation that because of the differences in net radiation gains and losses, energy must be transferred from the earth to the air and from low to high latitudes. The planetary winds, the depressions and the thunderstorms are all part of the exchange systems set up by these differences and organized to bring about the necessary transfers and balances. The atmosphere is rather like a complex heat engine in that differences in the net radiation balance supply the thermal power and that its various motions supply the means of distributing the energy. As in an engine, heat energy is converted into kinetic energy, the energy of motion.

Chapter 5

The atmosphere in motion

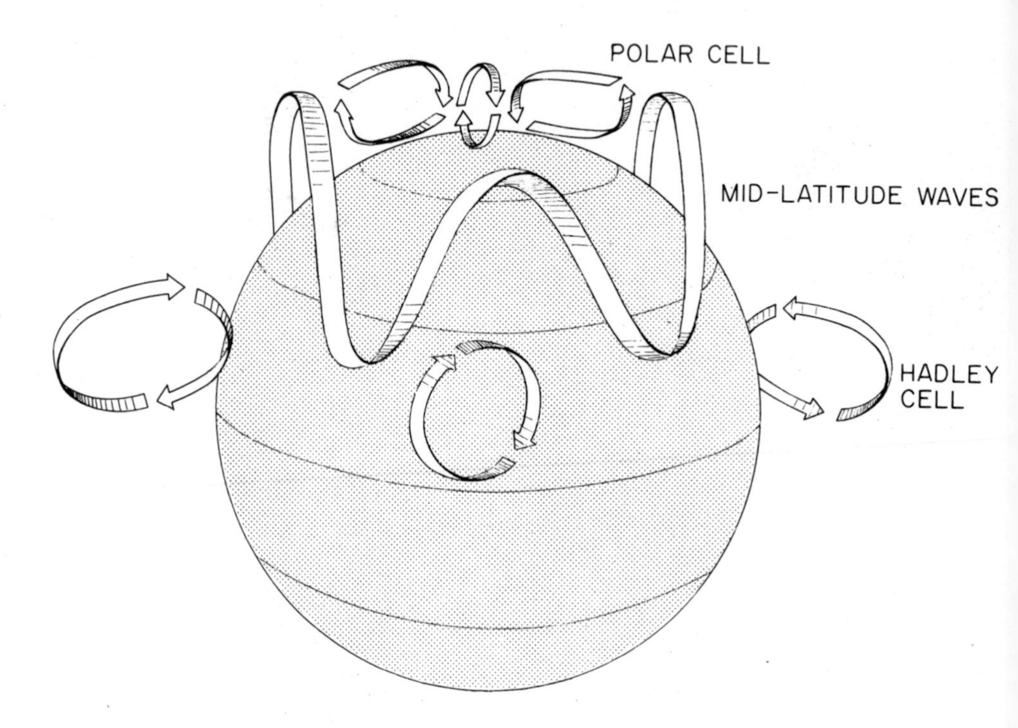

Fig. 6 The primary hemispheric circulation patterns.

The atmosphere is constantly in motion on scales varying from fractions of a millimetre to thousands of kilometres. The motion is produced by differences in energy distribution and is organized in the constant attempt to smooth out those differences.

On the largest scale, a great deal of scientific effort has been directed in recent years towards an improved understanding of the nature and cause of the average, long period patterns of the very basic, simple trends of atmospheric motion known as **general circulation.** Not only is this general circulation fundamental to the understanding of the patterns of climate but also, it provides the surest avenue to a satisfactory system of long-range forecasting. Research approaches vary from observation and theory to the simulation of conditions in laboratory (so-called 'dish-pan') experiments. Computers have made possible the complex calculations which were previously impossible and satellites have enabled us enormously to improve our knowledge of atmospheric conditions. In consequence we now have a much better idea of what is going on than we had even ten years ago.

Many models have been produced to portray the basic features of the general circulation but the essential form is shown in Fig. 6. Figs. 7 and 8 give more detailed, meridional pictures. There are, it will be seen, three main units of the circulation in each hemisphere, though their strength varies from day to day, between seasons and from one part of the world to another, even in the same latitude belt. One must remember that the general circulation represents only the basic essentials of air movements free from all the complications of individual depressions, anticyclones, monsoonal influences and the like.

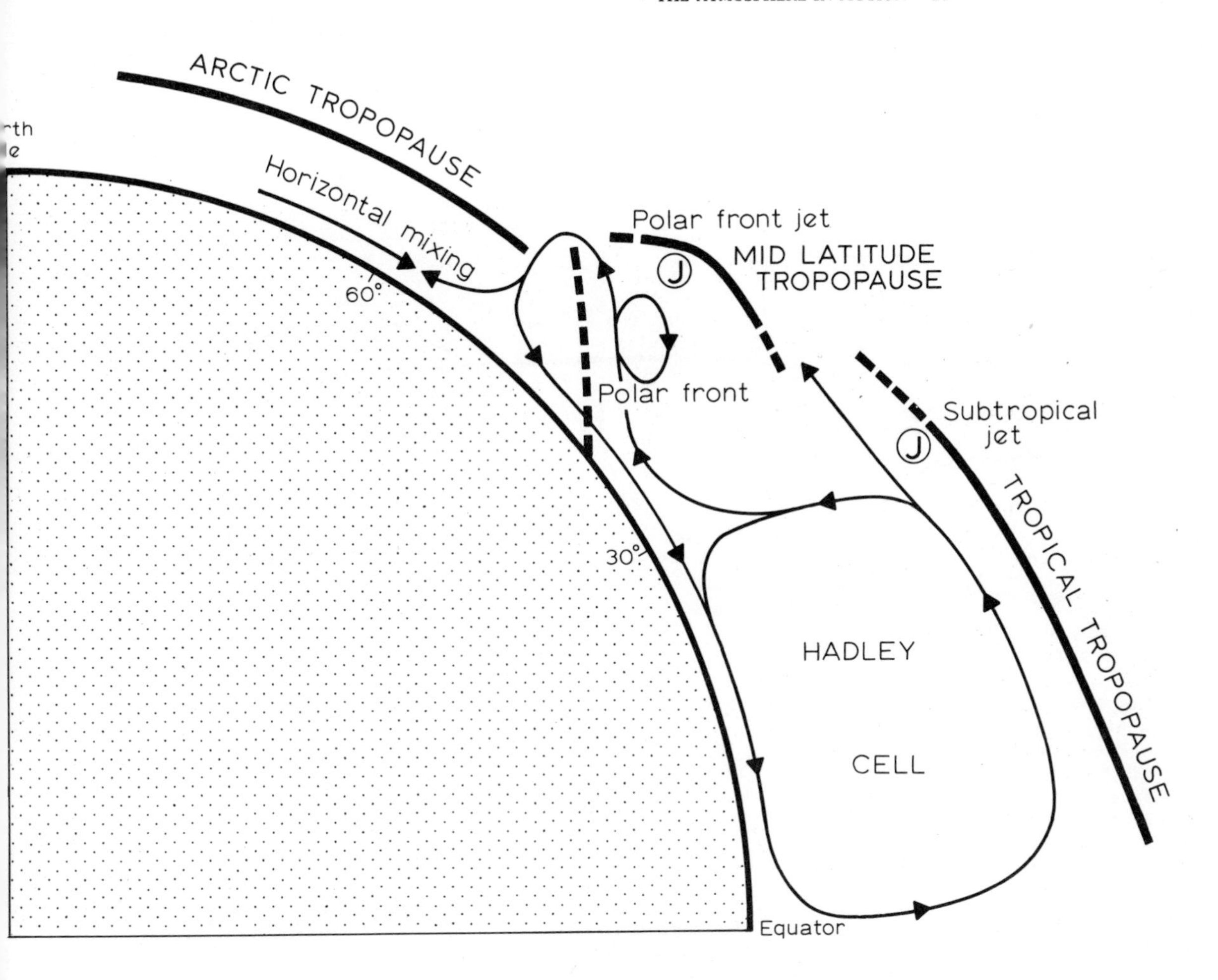

Fig. 7 General meridional trends of atmospheric motion in the northern hemisphere during winter. After diagram in E. Palmên, 'The Rôle of Atmospheric Disturbances in the General Circulation', *Quart. J. Roy. Met. S.*, Vol. 77, 1951.

The Hadley cell

In tropical latitudes, and more especially in the winter hemisphere, the atmospheric circulation is dominated by the meridional **Hadley cell** (Figs. 6 and 7), named after the Englishman George Hadley, who, in 1735 first proposed the idea of a tropical cellular circulation. In this cell, air near to the earth moves equatorwards as the Trade winds. These pick up moisture and thereby latent heat over the oceans and this heat is liberated when the vapour condenses in the towering convective clouds which occupy the ascending limb (a broad area of rising air) of the cell.

The heat liberated in the ascending air on the equatorwards side of each Hadley cell then drifts polewards at the top of the cells. At the same time, warm ocean currents in each hemisphere also help to carry sensible heat to higher latitudes: we need more observations before we can be sure of their importance but some estimates are that ocean currents are responsible for one fifth of the heat transfer from the tropics.

The Ferrel westerlies

In middle latitudes, air in the so-called **Ferrel westerlies** moves from west to east through a series of waves (Figs. 6 and 8). They are named after the nineteenth-century American meteorologist, William Ferrel. The largest of the waves, known as the **long or Rossby waves,** stretch from sub-polar to tropical latitudes and there are generally about four in each hemisphere (Fig. 9). They are a dynamic response to the thermal gradient between low and high latitudes on a rotating earth

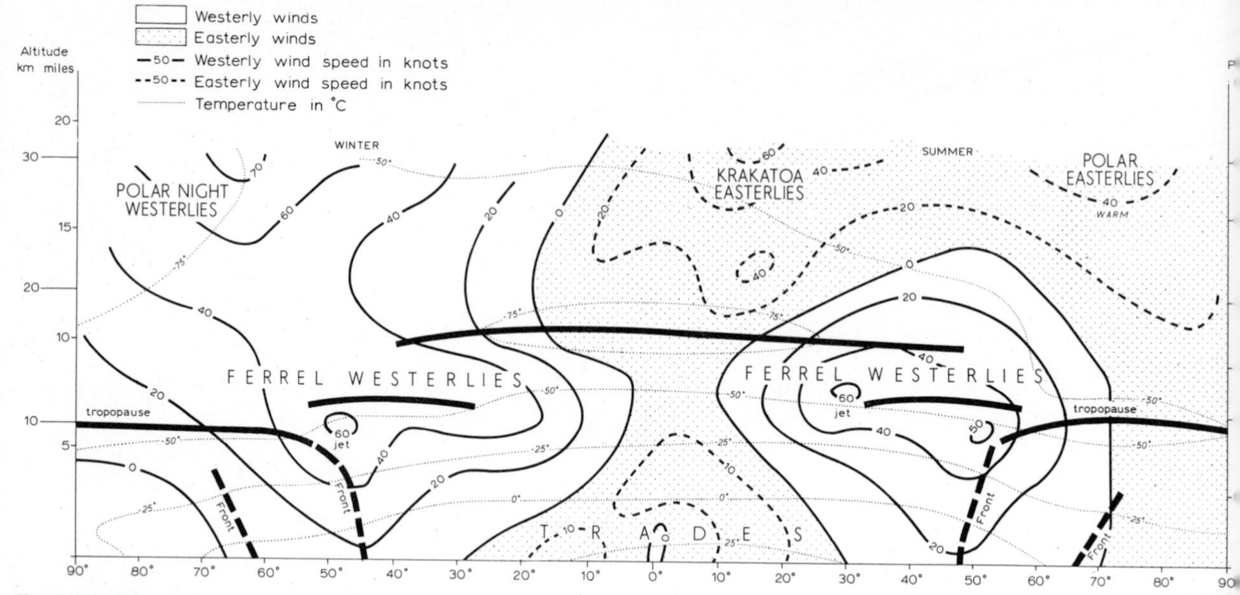

Fig. 8 Westerly and easterly winds and air temperatures in a meridional section through the troposphere and lower stratosphere along a line running down the central Atlantic. After diagram in H. H. Lamb, *Geography*, Vol. 46 (3), 1966.

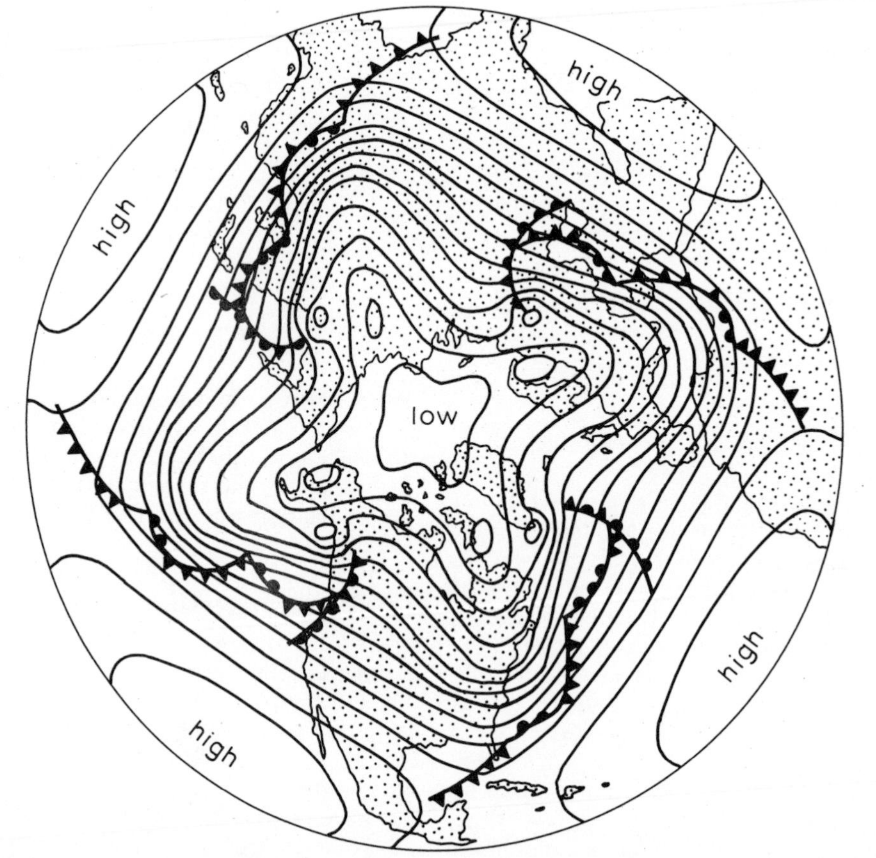

Fig. 9 Polar view of mid-tropospheric (altitude about 5 km) pressure waves and surface fronts around the northern hemisphere. After diagram in E. Palmên, 'The Aerology of Extra-tropical Disturbances', *Compendium of Meteorology* (American Meteorological Society, Boston, Mass., 1951).

and to air rising and falling over the Western Cordillera and Andes mountains. Their positioning is influenced by the seasonal temperature differences between land and sea. The reasons why winds meander downstream of a major topographic barrier are summarized in Fig. 10. Over the crest, the air contracts vertically

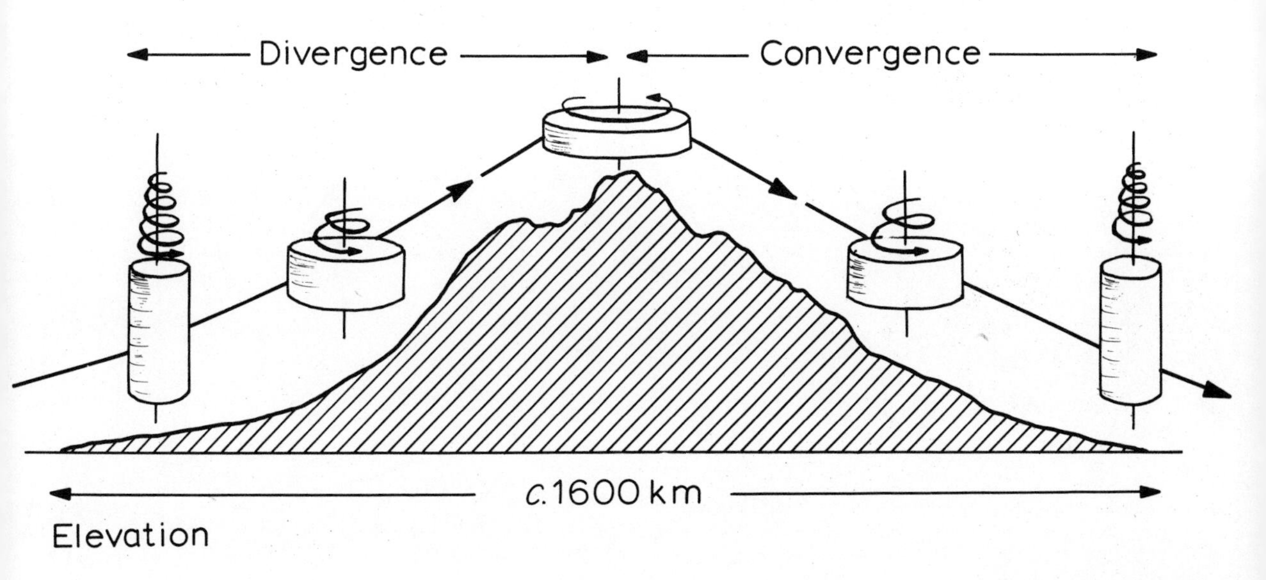

Fig. 10 The effect of a broad, high topographic barrier such as the Western Cordillera or Andes upon a westerly airstream.

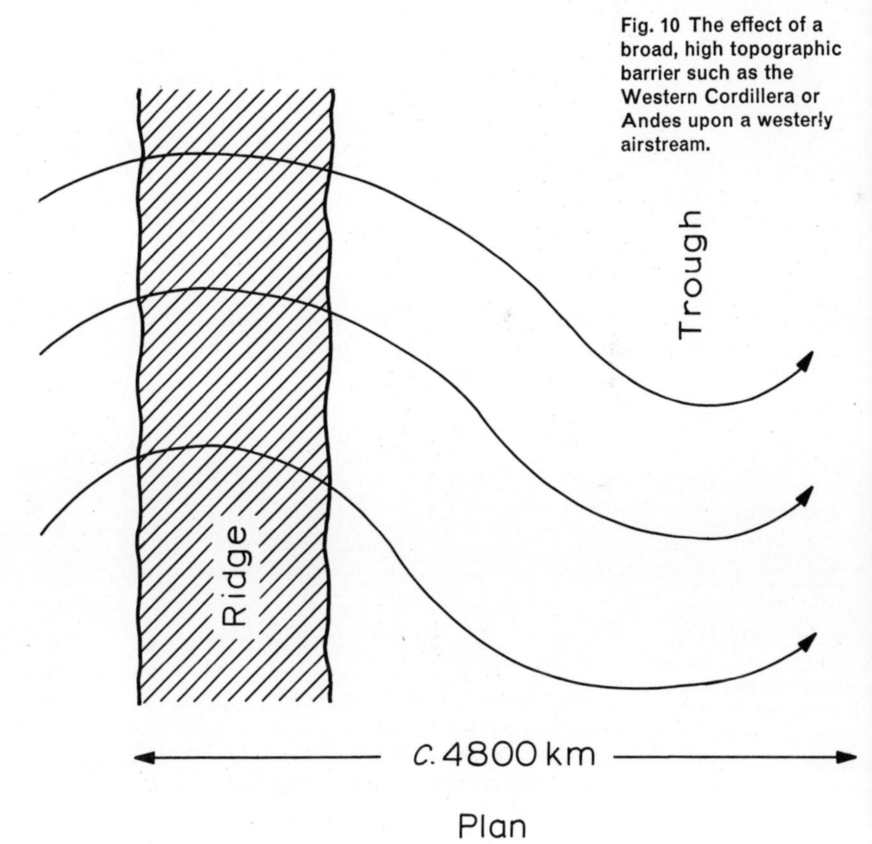

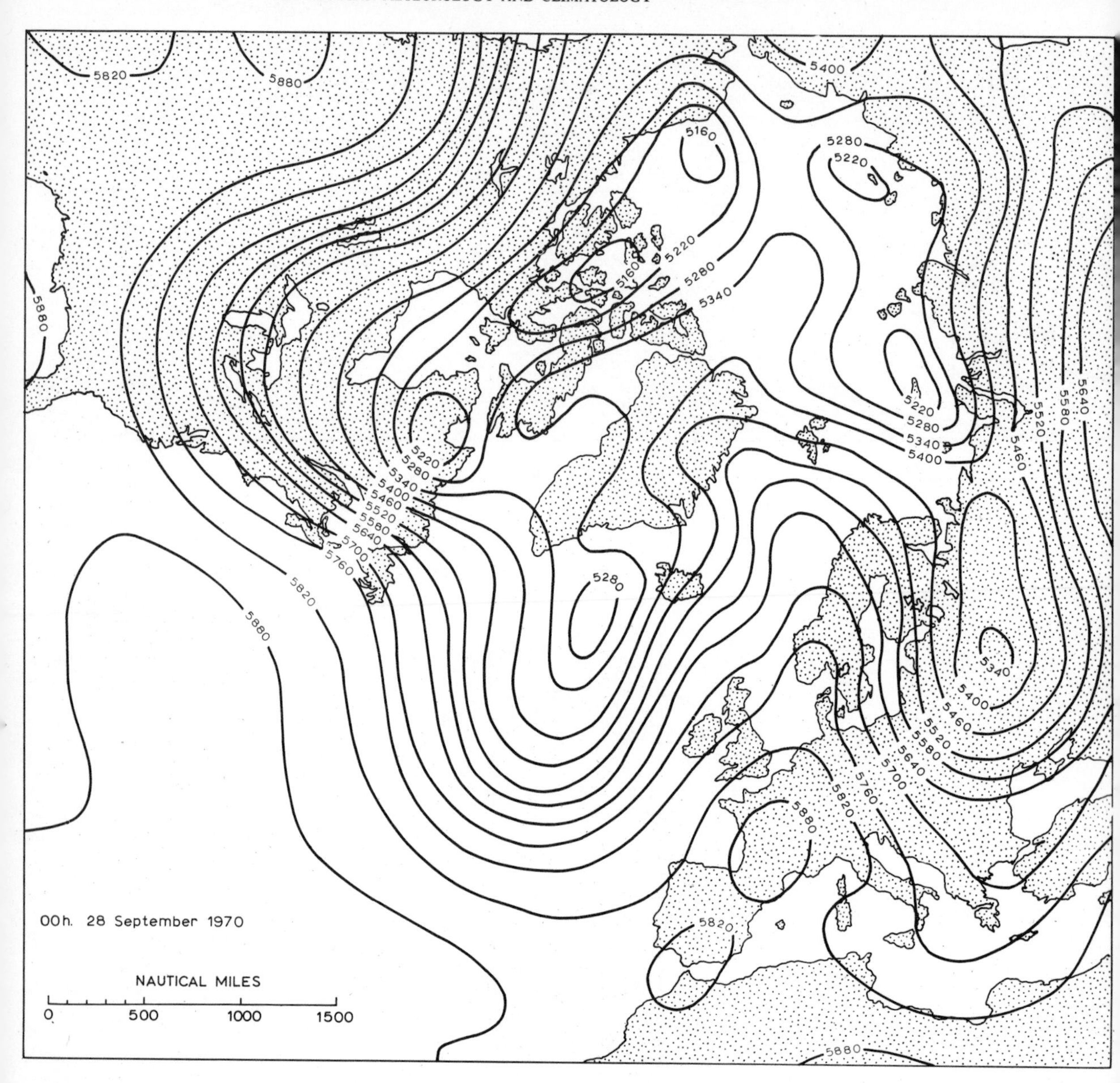

Fig. 11 Contours (in metres) of the 500 mb pressure surface above part of the northern hemisphere. 00h, 28 September, 1970.

and diverges horizontally, as a result of which its rotation relative to the earth (relative vorticity) is reduced in much the same way as an ice skater revolves more slowly when she crouches down and spreads out her arms. This is because angular momentum (mass × velocity × radius) is conserved and an increase in radius is balanced by a decrease in velocity. As the air descends the leeward slopes of the mountains, there is horizontal convergence and the column of air is stretched vertically so that there is an increased rate of spin in the same direction as the earth. Over the crest of the mountains there are therefore increased anticyclonic tendencies and over the leeward slopes, the air moves cyclonically. In this way waves are established with anticyclonic curvature at their crests and cyclonic curvature in their troughs. Because of the earth's rotation and the air's movement latitudinally these waves are perpetuated downstream, assisted perhaps by further topo-

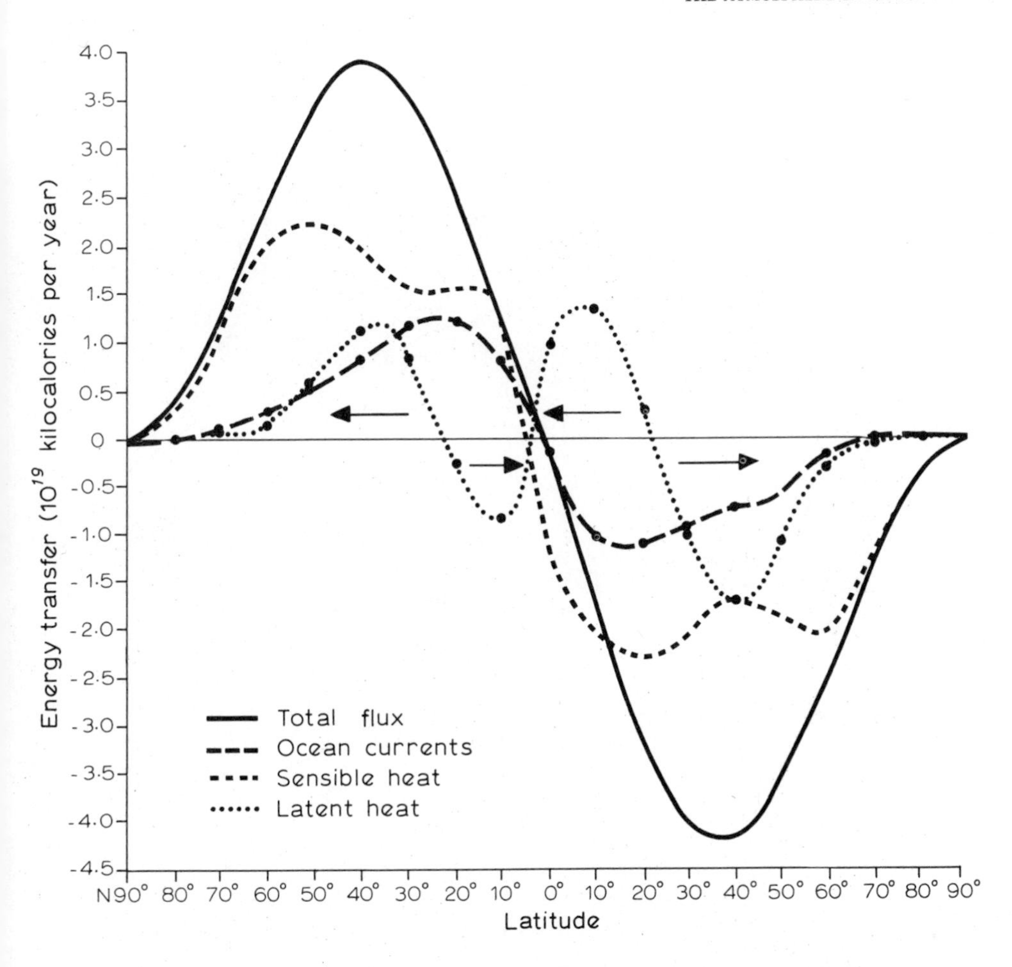

Fig. 12 The latitudinal variation in average meridional energy transfers in the earth–atmosphere system. After diagram in W. D. Sellers, *Physical Climatology* (University of Chicago Press, 1965).

graphic barriers such as the Tibetan Plateau.

Because temperatures are lower over the Antarctic than over the Arctic, the intensities of the southern hemisphere westerlies are much greater than those in the northern hemisphere and they also extend nearer the Equator.

Moving through these pseudo-stationary long waves are shorter waves set up by instabilities in an airstream across which there is a moderate thermal gradient. Like the long waves, they serve to transport heat from side to side and as they move through the long waves in the middle and upper troposphere (Fig. 9) a fairly complex overall picture is produced (Fig. 11). In general, cold air sinks in the western limb and warm air rises in the downstream or eastern limb of these waves.

In polar regions there is a weak cellular overturning known as the polar cell (Fig. 6) but this frequently disappears to give low pressure over the poles.

Through these various links in the general circulation, set up by radiation on a rotating earth producing thermal and pressure differences, energy is transferred in a variety of forms from areas of surplus to areas of deficit (Fig. 12). In so doing, the fundamental character of the world's patterns of winds and climates are established.

Low latitude circulations

The Hadley cell establishes a pattern of pressure and a system of winds that goes a long way towards explaining the weather of low latitudes (Figs. 13 and 14). Except over the central Pacific, the rising

Fig. 13 The global distribution of pressure and winds, January.

Fig. 14 The global distribution of pressure and winds, July.

7. This photograph, taken on 14 June, 1967, from a Nimbus satellite, shows the distribution of cloud during anticyclonic conditions over the British Isles. There is stratus cloud over the North Sea and frontal cloud banks are decaying over Ireland.

limbs of the Hadley cells generally lie on either side of a belt of low pressure known as the **doldrums** which for most of the year produces calms or light winds broken by thunderstorms. In the central Pacific and temporarily elsewhere, one finds the **intertropical convergence zone** where the Trade winds from the two hemispheres are sharply drawn together in a zone of rising air. This convergence zone migrates north and south with the seasons with a time lag of up to one month behind the vertical sun. It travels further over the land than over the sea because, for several reasons, land warms faster than water.

In the poleward limb of the Hadley cell, sinking air produces the high pressure cells (anticyclones) of the so-called **horse latitudes.** Easterly Trade winds blow around the equatorward side of these anticyclones and on their poleward side the westerly winds form part of the mid-latitude westerlies. Except over the Atlantic Basin these simple patterns of air movement are sharply disturbed by the monsoonal reversals of airflow which typify the Indian and Pacific Oceans.

Anticyclonic subsidence leads, of course, to the compression and warming of the air, but this can occur only down to the upper limit of strong convection rising from the warm earth. There is therefore no warming below about 6000 to 10 000 ft (1800 to 3000 m) in the tropical anticyclones and a sharp temperature maximum known as an **inversion of temperature** (see p. 33) forms at this height. It is called the **Trade wind inversion** and it frequently damps down or halts currents of air rising from the ground. In consequence, clouds

8 .By comparison a satellite photograph taken at 6 a.m. on 21 January, 1971 shows a deep depression centred just SW of the British Isles with the occlusion running NW–SE across England.

are often stunted and only occasionally give rain. This is the fundamental explanation of the tropical (Trade wind) deserts.

Mid-latitude circulations

In mid-latitudes, weather is dominated by the consequences of air movement through the stationary and migratory waves of the Ferrel westerlies in the middle and upper troposphere. The ridges and troughs of these waves are linked to surface anticyclones and depressions. The very large, seasonally almost stationary, anticyclonic and cyclonic areas are the surface representation of the major, long waves in the upper westerlies while the migratory highs and lows are linked to smaller waves. As air moves round the sharp curves of the latter, it is subject to accelerations and decelerations as the apparent centrifugal force alternately reverses direction, first one way and then the other, around successive crests and troughs. In consequence of this and the more rapid forward motion of waves in the middle troposphere than near the ground, the air in the middle and upper troposphere is divergent and picks up anticyclonic spin ahead of a trough and is convergent with cyclonic spin behind the trough (Fig. 15). Near the ground, where the wave pattern moves faster than the wind, there is compensatory convergence and cyclonic spin ahead of the trough and divergence and anticyclonic spin behind, so that the surface depressions and anticyclones and upper waves in the westerlies are typically related as shown in Fig. 15.

The near-surface convergence leads to a concentration of the isotherms between

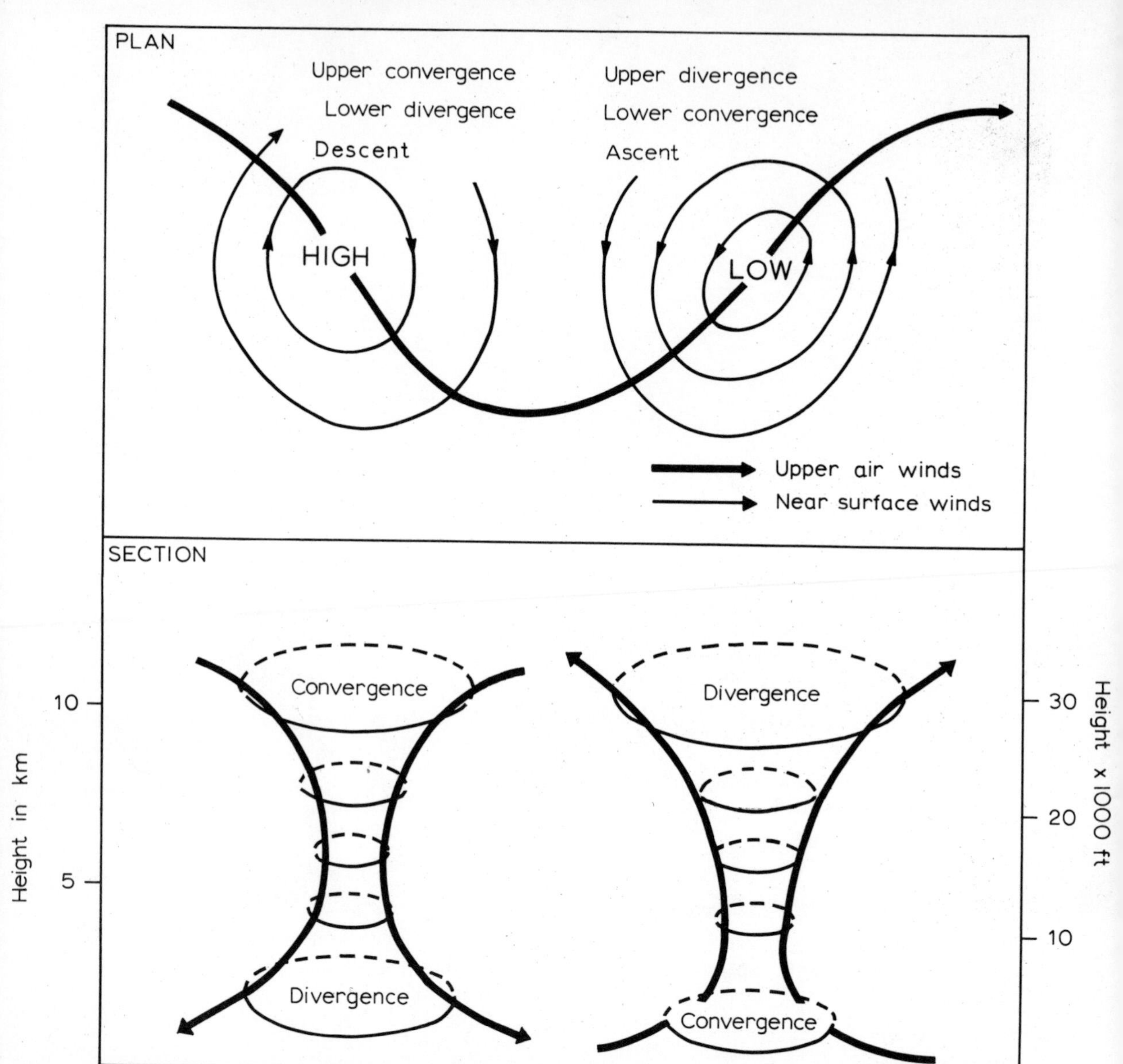

Fig. 15 Horizontal winds and vertical air movements in association with a wave trough (indicated by an upper air wind arrow) in the upper westerlies. After diagram in H. Riehl, *Introduction to the Atmosphere* (McGraw-Hill, 1965).

cold north-westerly airstreams and warm south-westerly air, that is, to **fronts.** Because of the convergence, there is also vertical motion with condensation and precipitation. Also, as surface winds climb above the earth they accelerate and their speed parallel to the front is further increased by a **thermal wind** factor developed by the strong thermal gradient which produces a pressure gradient across the front. This is because pressure falls more rapidly with height in cold, dense air than in warm, light air (Fig. 16). Above the tropopause, the direction of the thermal gradient is reversed (at the level of the tropopause, temperatures are lowest above the Equator) and so the pressure gradient is reduced and the wind rapidly decelerates. Near to the front at the level of the tropopause we therefore find a wind maximum known as a **'jet stream'** with winds up to 300 m.p.h. (500 km/h). There are frequently two jets in each hemisphere: one related to the discontinuous polar front lying in mid-latitudes where the thermal gradient between polar and tropical air has been intensified by novements in the mid-latitude waves, and a sub-tropical jet lying

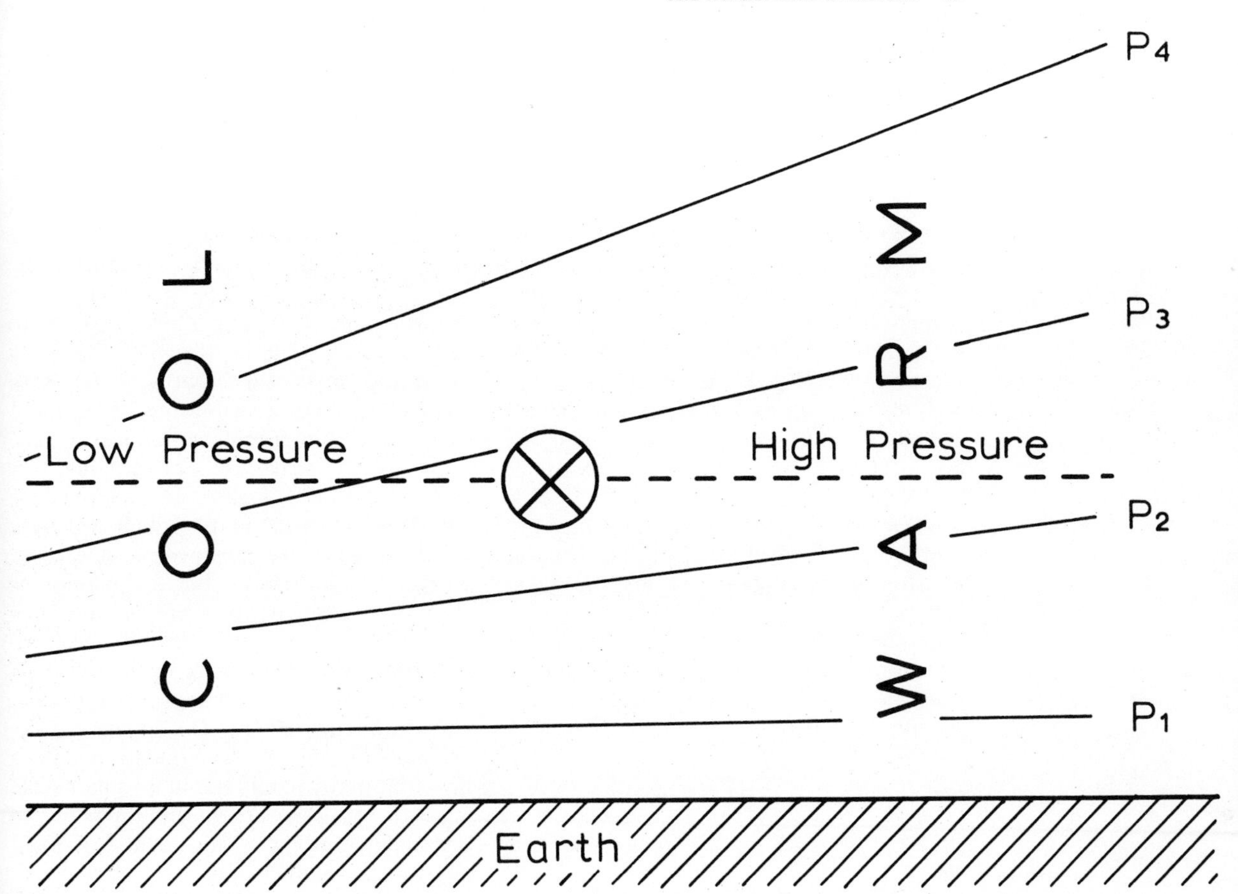

Fig. 16 Distortion of pressure surfaces and the creation of a thermal wind (into the plane of section) by a transverse thermal gradient.

on the poleward side of the Hadley cell in the upper troposphere (Figs. 7 and 8). The sub-tropical jet is more continuous around each hemisphere than the polar front jet but unlike the latter, it is not normally related to surface fronts.

The features of a mid-latitude depression are thus explained more realistically without resorting to the concept of air masses as in previous theory (*see* Chapter 7 for further details).

Polar circulations

In high latitudes where the troposphere is only half as deep as in the tropics, the circulation systems are shallow and often short lived. As in the tropics, weather in the polar regions is by no means as constant as was once thought although at present we know far less about the atmosphere of both these regions than we do about mid-latitude climates. Fortunately, co-ordinated efforts are now being made by many countries to improve our knowledge of the meteorology of high and low latitudes.

Winds are thus a consequence of inequalities in net radiation receipts and at the same time a means of balancing them out by transfers of energy in various forms. Energy is transferred both vertically and latitudinally in the Hadley cell. The depressions and waves of middle latitudes function in the same way and in latitude 38°, where there is the greatest latitudinal transport of energy, sensible and latent heat transfers are almost equal. In higher latitudes, most of the energy is transported as sensible heat (Fig. 12).

The water balance

As with energy, there is a roughly constant amount of moisture in the atmosphere in spite of local inequalities of evaporation and precipitation. And like heat, moisture is carried from source regions where evaporation exceeds precipitation, to sink areas where the reverse is true. At present we are unable to calculate the moisture balance for large areas of the globe because of the sparsity of precipitation data and the extreme difficulty in measuring or accurately calculating evaporation. Reasonable estimates can, however, be made of evaporation using other observations such as temperatures and windspeeds and in this way we can demonstrate the very low evaporation rates in polar regions because of the low temperatures and frequent calms, and the enormous water losses, often amounting to more than 100 inches per annum (254 cm/yr), from the areas of warm seas and strong winds in the western Atlantic and Pacific.

Conclusion

The problems of the general circulation are being tackled in a number of different ways. Using the largest of modern computers, theoretical models are being formulated and tested against our increased observational knowledge which, in recent years, has been greatly enhanced by meteorological satellites (Plate 6). The models are becoming more and more sophisticated, that is, they are including more and more real features of the earth such as the basic distributions of land and the major topographical features of the earth. The results are encouraging, for already by these means we can produce simulated climatic maps which are very close to what we observe. Much more work needs to be done but this is perhaps the most exciting period there has ever been in meteorology for it is clear that for the first time, in detail, we are beginning truly to understand the complex, three-dimensional workings of the atmosphere.

By definition, general circulation studies filter out many medium and all small-scale features and these are vitally important to the understanding of the details of local day-to-day weather. The next section will look at some of these processes.

Chapter 6

Evaporation, condensation and precipitation

It is sometimes said that cloud forms are the writings of the atmosphere, but to understand individual messages it is necessary to learn the meteorological language and to be able to distinguish between indicators of major processes and minor local trimmings. Clouds are one link in a complex cycle of moisture between and within the earth and its atmosphere. Reducing these links to their fundamental forms and neglecting the many feed-backs and short circuits, the linked sequences are known as the **hydrological cycle,** the essential features of which are shown in Fig. 17. The great majority of clouds evaporate rather than precipitate but eventually their moisture will be returned to the earth only to be evaporated and then condensed again in a new cycle.

The hydrological cycle operates over a great variety of time scales and space scales but on average it takes about 10 days to complete, during which time there may be several thousand miles separating the areas of evaporation (including transpiration) and precipitation, and the associated abstraction and release of latent heat respectively.

The amount of moisture in the atmosphere at any one time is about 14×10^{12} (14 million million) tons, which is the equivalent of 1 inch (25.5 mm) of rain over the entire earth. The vast majority of the atmosphere's moisture is of course held as an invisible gas—water vapour—and because warm air can hold proportionally much more of this gas than colder air, some of the greatest quantities of moisture drift invisibly above deserts. The problem of aridity is therefore of condensation and precipitation rather than any lack of water vapour.

Moisture exists in the atmosphere in all three states of matter: gaseous, as invisible water vapour; liquid, as clouds, fog and rain droplets; and solid, as ice crystals and hailstones. Changes from one state to another can only occur in certain circumstances, a knowledge of which is vital to the understanding of most weather processes.

Evaporation

Because we can detect no long-period trends in the amount of moisture in the atmosphere, it follows that evaporation into the atmosphere must be balanced by its removal through condensation and precipitation. But it is very difficult to measure evaporation locally since it occurs invisibly from most natural surfaces varying from seas and lakes to soils, leaves and bricks. The instrument which comes closest to directly measuring evaporation and transpiration from vegetation-covered soils is known as a lysimeter and works by isolating a section of soil with its vegetation and measuring the rainfall, the change in the amount of water in the soil and the percolation, so that the evaporation can be calculated as the fourth term in the simple equation:

$$Pr = E + S' + Pe$$

where Pr is the precipitation, E is the evaporation, S′ is the change in stored water in the soil, and Pe is the percolation through the soil. These instruments are very

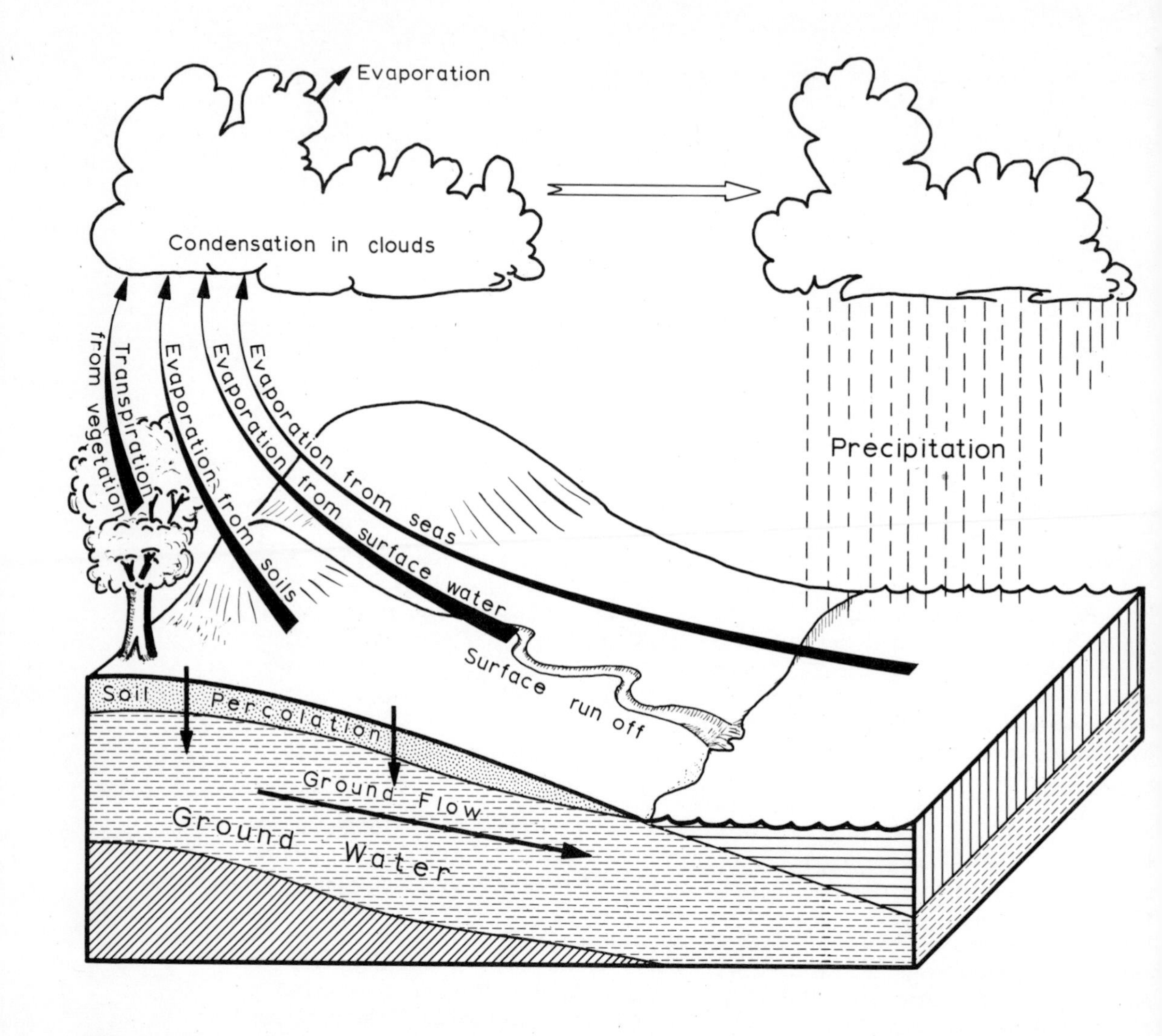

Fig. 17 The hydrological cycle linking evaporation (and transpiration), condensation, precipitation, surface and subsurface flow.

costly and difficult to use and they are designed to measure evaporation from crop and grass-covered soils, not being very suitable for use with trees and most other natural surfaces. Meteorologists have therefore devised formulae for calculating rather than directly measuring evaporation and among the best known of these is the one by the late C. W. Thornthwaite, an American geographer.

As a consequence of these studies, we now have more detailed figures for evaporation rates, and meteorologists have been able to apply their knowledge to such problems as the design and operation of irrigation systems and reservoirs and the effect of forests upon water yield in catchment areas.

Cooling and saturation

The warmer the air, the more water vapour it can hold (Fig. 18). When it holds the maximum amount at a given temperature, the air is said to be **saturated** and the partial pressure of the water vapour among all the other gases is known as the **vapour pressure**; in saturated air it is known as the **saturated vapour pressure.**

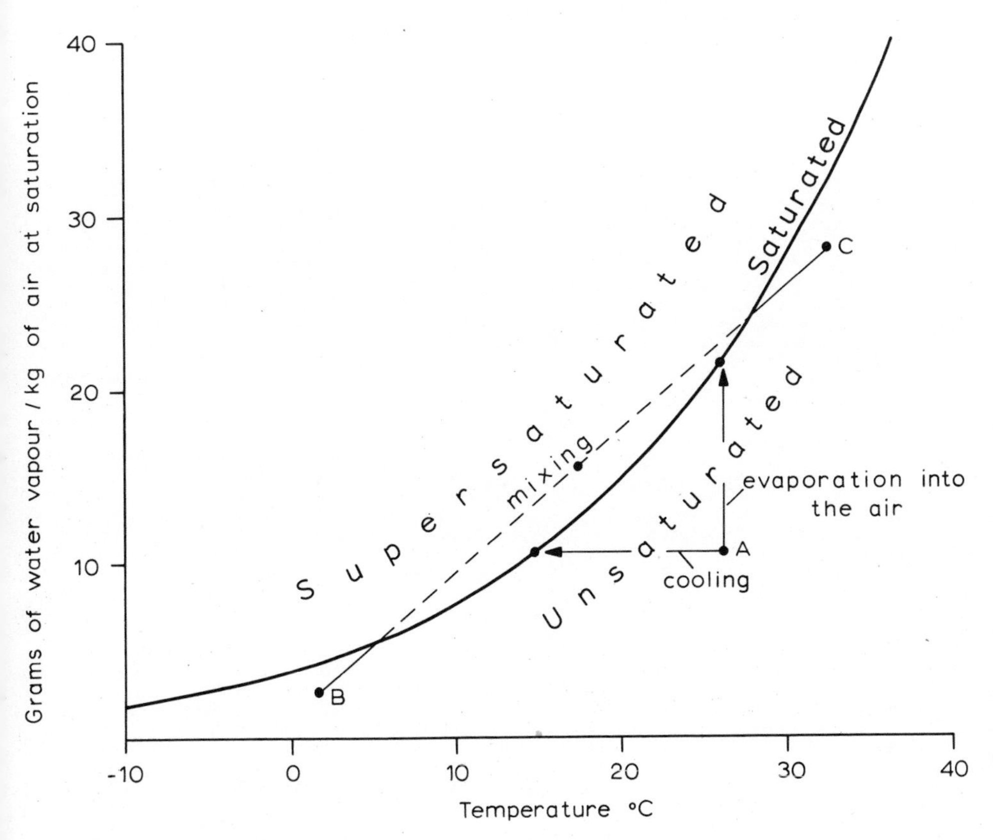

Fig. 18 The weight of water vapour in saturated air at different temperatures. An unsaturated sample of air, A, can be brought to saturation either by increasing the moisture content and/or by cooling. If air samples B and C are mixed together they will form a supersaturated mixture.

The ratio between the vapour pressure and the saturated vapour pressure is known as the **relative humidity.** Saturated air has a relative humidity of 100 per cent and near to the earth, the ratio very rarely falls below 30 per cent. Water vapour weighs less than other atmospheric gases so that moist air weighs less than dry air and warm, moist air weighs least of all.

Suppose we now consider a parcel of unsaturated air A, whose temperature and vapour content are indicated in Fig. 18. There are then an infinite number of ways of changing the water vapour content and temperature of the air to make it saturated. One method would be to keep the temperature constant and to increase the vapour content and another to keep the vapour content constant and to decrease the temperature. Another possibility is for two masses of air, each not quite saturated, to mix together. The latter process may happen along frontal zones (see p. 24) although it probably explains only a small proportion of clouds. Cooling is quite definitely the main process by which moist air is brought to saturation in the atmosphere. The cooling can be achieved by advection (that is, horizontal movement) over a cool land or sea surface, or by vertical, convective-type movements. The former produces fog or low cloud and the latter gives cumulus and stratocumulus types of cloud. Overall, vertical movement is far more important than horizontal movement in producing cloud, and air rises either because of its natural buoyancy which stems from its warmth and humidity, or through forced ascent over an upland area or a mass of cold air of higher density.

The buoyancy of the air

Temperature generally and pressure invariably falls with height. Rising air therefore expands and cools and there is also some mixing with the surrounding cooler air. The rising air is often organized as a **thermal** (Fig. 19) around which the air is sinking and warming.

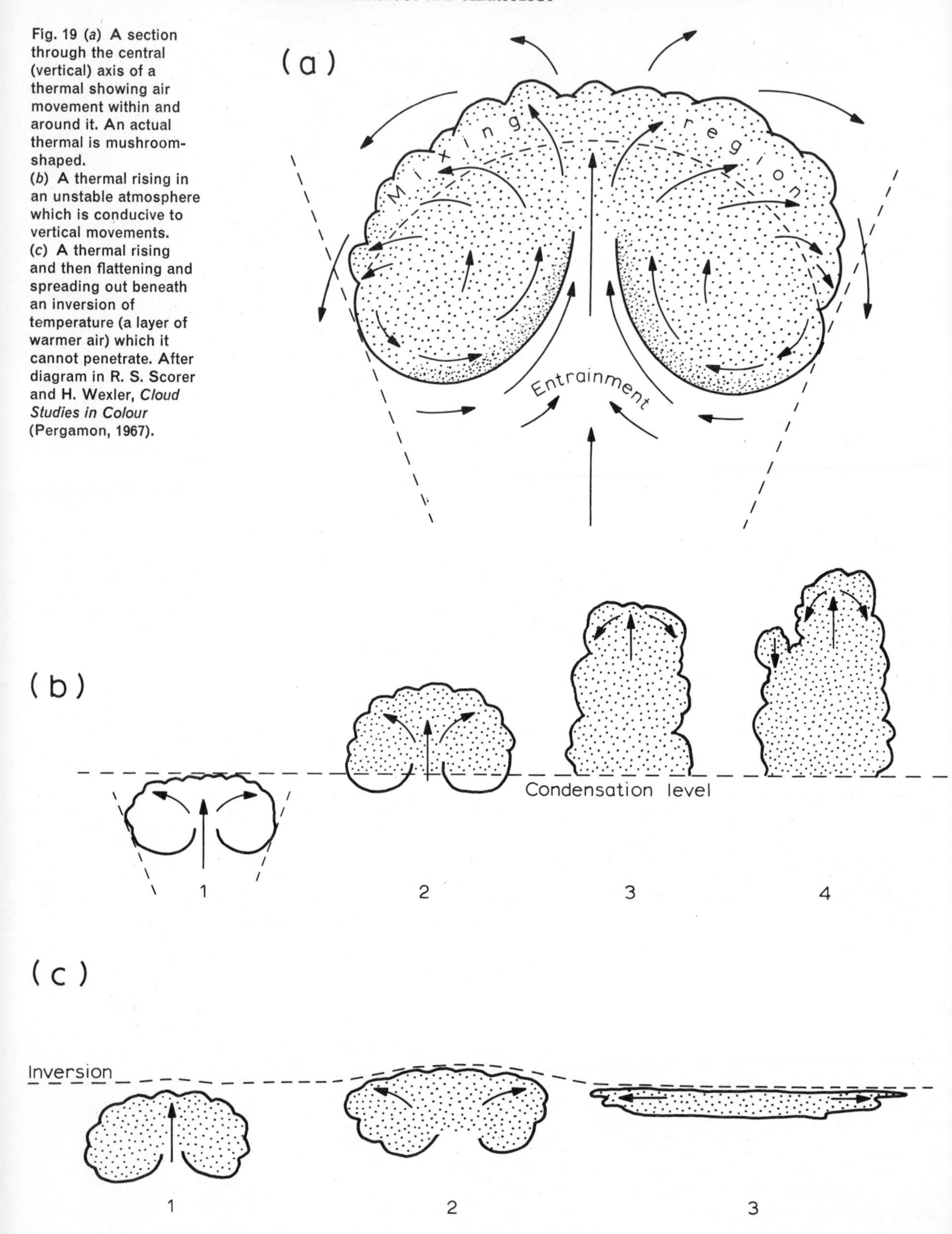

Fig. 19 (*a*) A section through the central (vertical) axis of a thermal showing air movement within and around it. An actual thermal is mushroom-shaped.
(*b*) A thermal rising in an unstable atmosphere which is conducive to vertical movements.
(*c*) A thermal rising and then flattening and spreading out beneath an inversion of temperature (a layer of warmer air) which it cannot penetrate. After diagram in R. S. Scorer and H. Wexler, *Cloud Studies in Colour* (Pergamon, 1967).

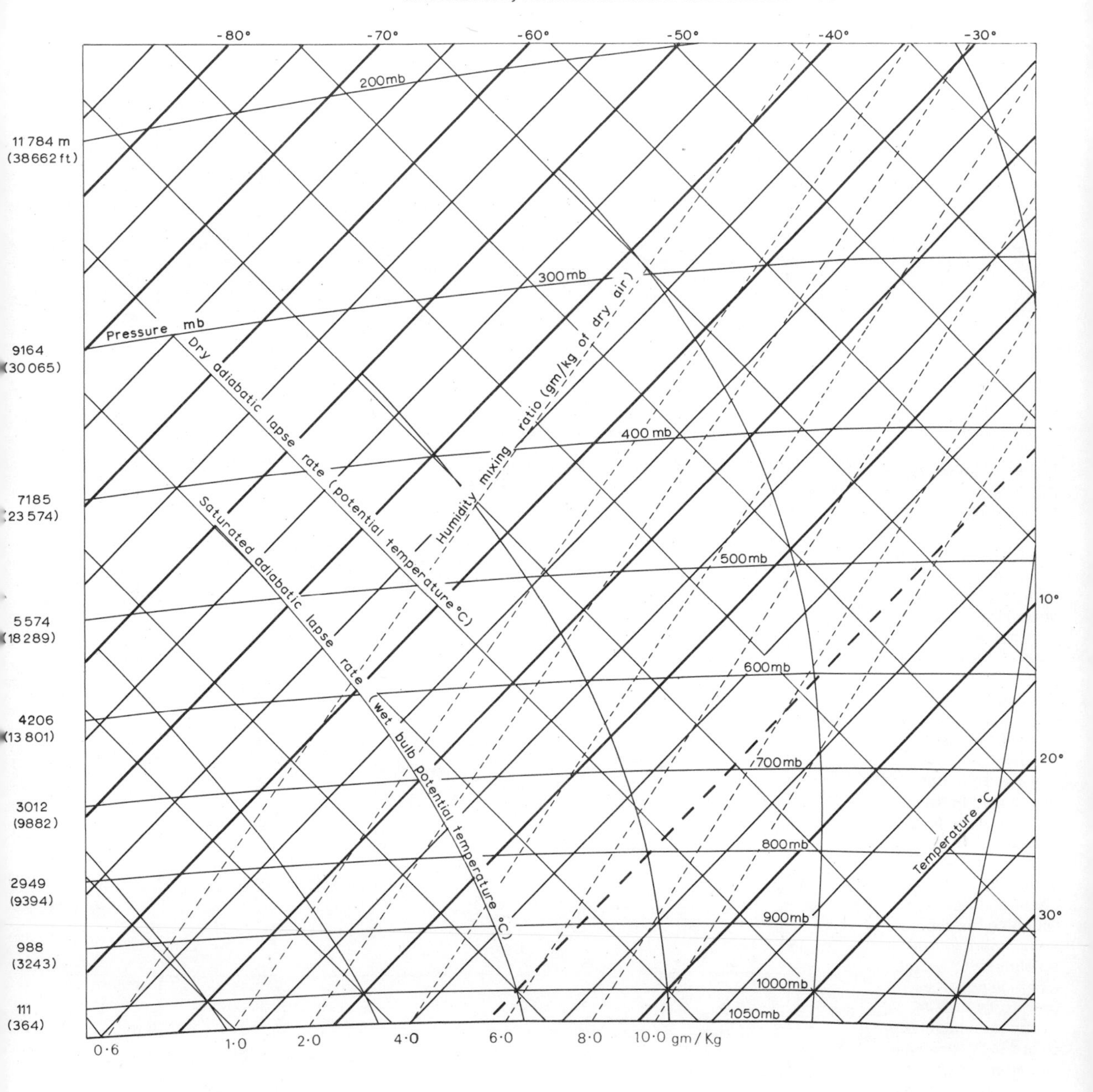

If the air rises swiftly, as it often does, the amount of mixing is small. In this case we can regard the parcel as having no energy exchanges with its environment by either conduction or convection although there will be important conversions of energy within the ascending air. More particularly heat energy will be converted into kinetic energy (used in expansion) and

Fig. 20 A tephigram, used to plot air temperature against height (pressure) and to calculate the stability of the atmosphere and the height and depth of clouds. The height of the cloud base can be determined by noting where the dry adiabatic lapse rate line from the surface dry bulb temperature intersects the saturated adiabatic lapse rate line from the surface wet bulb temperature. Another method is to note where the dry adiabatic lapse rate line from the surface dry bulb temperature intersects the humidity mixing ratio line equivalent to the measured humidity mixing ratio (grams of water vapour per kg of dry air) at the surface.

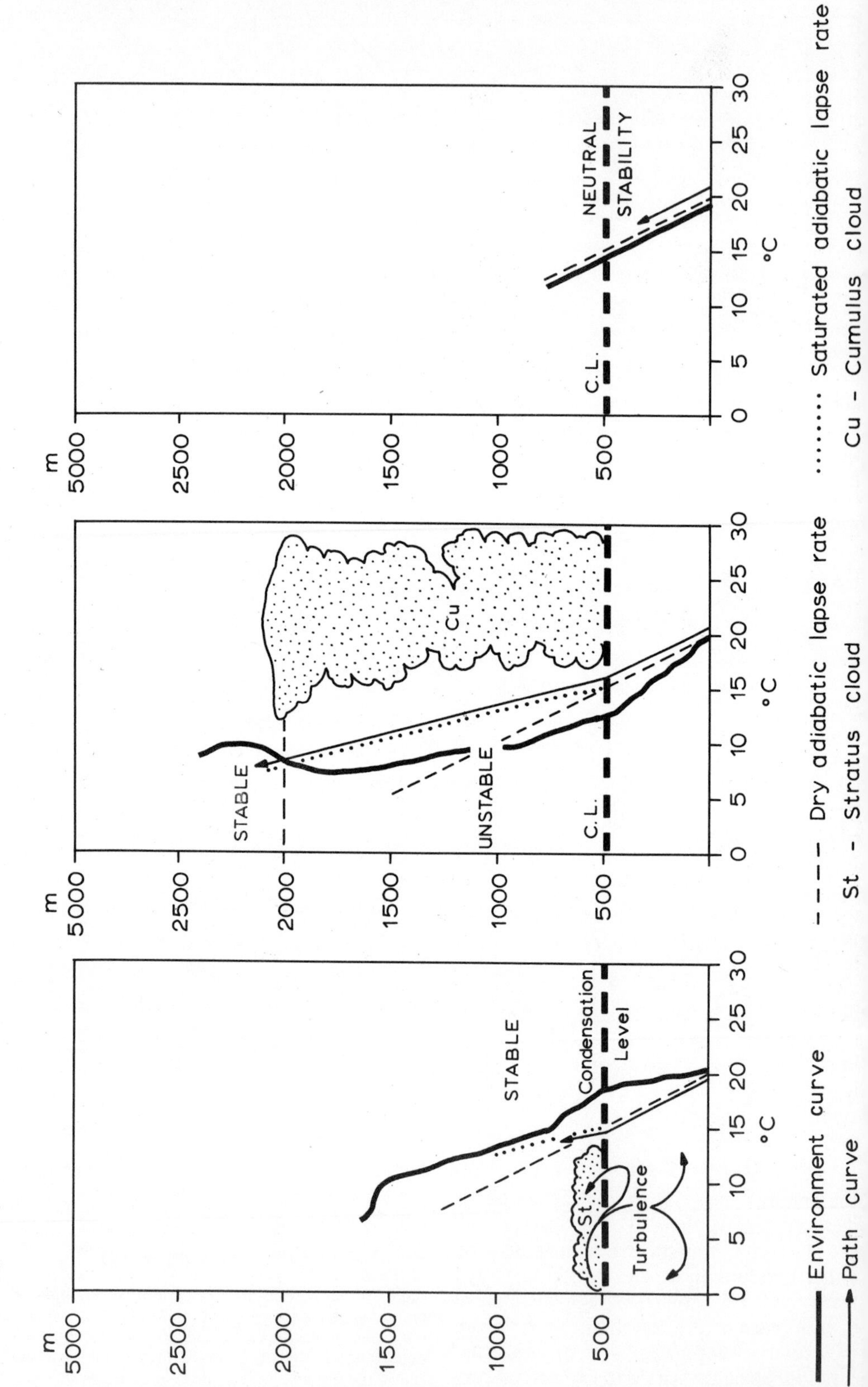

Fig. 21 Graphs showing the relationships between the temperature of air rising adiabatically (the path curve) and the temperature of the surrounding air (the environment curve) in stable, unstable and neutral atmospheres. Air rising from the ground will cool dry adiabatically until condensation takes place at and above the condensation level. With further ascent it will cool at the saturated adiabatic lapse rate. When the rising air is cooler than the surrounding air, the atmosphere is said to be stable and air will sink unless forced upward by turbulence, by topographic barriers or at fronts. Unstable atmospheres are conducive to uplift because the rising air is warmer and thereby lighter than the air around it but a higher, stable layer may limit the uplift and flatten the tops of clouds. With neutral stability, the air temperature of the rising air is the same as that of the surrounding air and in the absence of other forces it will neither rise nor sink.

potential energy (because its centre of gravity is raised). For these reasons, the parcel will lose heat energy and will cool. In the absence of condensation, the air will cool at the **dry adiabatic lapse rate** (DALR) of 1°C/100 m (1°C/325 ft). If there is condensation, latent heat is released as the vapour condenses into droplets and this added warmth partly compensates for the cooling owing to expansion so that the overall fall in temperature is less than for dry air. At high altitudes however, the air is much drier and the amount of latent heat released is only small. The rate of cooling of saturated air, known as the **saturated adiabatic lapse rate** (SALR) is therefore about half the dry adiabatic lapse rate near the ground but increases to almost equal it near the top of the troposphere. On **tephigrams** (Fig. 20), used to calculate the temperature and buoyancy of rising and sinking air, DALRs are therefore represented by straight parallel lines while SALRs are non-parallel convex curves.

The buoyancy of moving air depends upon its density in relation to that of the surrounding air. As density is inversely related to temperature, air that is warmer than its surroundings will rise while air that is colder will sink. If the lapse rate of the air (the variable rate at which temperature falls with height) is slack, that is the environmental temperature falls only slowly or even rises with height, then air rising into such an environment will be cooler and sinking air will be warmer than the surrounding air. Either way, the air will tend to return to its original position and vertical motion will be damped down. The air is said to be **stable** (Fig. 21). If, on the other hand, temperature falls rapidly with height in the environmental air, faster, that is, than the dry adiabatic lapserate, then rising parcels of air will become progressively warmer than their surroundings and their buoyancy will accelerate their rate of uplift. Such an atmosphere is said to be **unstable** and encourages uplifted air to rise and sinking air to sink even further. If the observed lapse rate equals the adiabatic lapse rate (which it will do only over short distances) the moving parcel will have the same temperature as its environment and it will remain where it is as long as buoyancy is the only control. Such an atmosphere is said to have **neutral stability.** The well-stirred atmosphere of the **boundary layer,** that is the lower 2000 ft (600 m) or so of air, is often in neutral stability but on hot, calm summer days, air near the ground will be very warm and the lower atmosphere will be very unstable with the possibility of thunderstorms. On calm clear nights, on the other hand, air near the ground will be cooled by radiating and conducting its heat to the ground cooled by long-wave radiation, and temperatures will increase with height in the lower 10 to 1000 ft (3 to 350 m) of air, a condition known as an **inversion of temperature.** Such inversions occur on about two nights in five in Great Britain and they are associated with **radiation fogs,** trapped near the ground beneath the warm air of the inversion. An inversion is also formed when warm air flows over a cold land or sea surface. If the cooling is sufficient to bring about condensation, then an **advection fog** will be formed. These are common when maritime air moves over land in winter and over cool coastal waters in spring and summer. Over cities, pollution is often mixed with the fog to produce a thick **smog** (Plate 9) which might prevent a great deal of radiation reaching the ground and this helps to perpetuate the condition.

Inversions above the ground can lead to layers of **stratus** cloud (Plate 10) or to the flattening of clouds rising from below, called **fair weather cumulus** (Plate 11). The **trade wind inversion** at heights of 6000 to 10000 ft (2000 to 3000 m) in the eastern parts especially of the tropical high pressure cells is largely responsible for the shallowness of the cloud layer and the aridity of these areas. On a larger scale again the stratosphere is extremely stable and flattens the tops of thunderclouds known as **cumulonimbus** clouds (Plate 12).

9. Heavy smog over Manhattan, New York.

10. *(Opposite page)* Low stratus cloud clinging to the mountains at Glascurrock, Scotland.

11. Fair weather cumulus at Glengarriff, Scotland.

12. *(Opposite page)* Cumulonimbus head (thunderclouds).

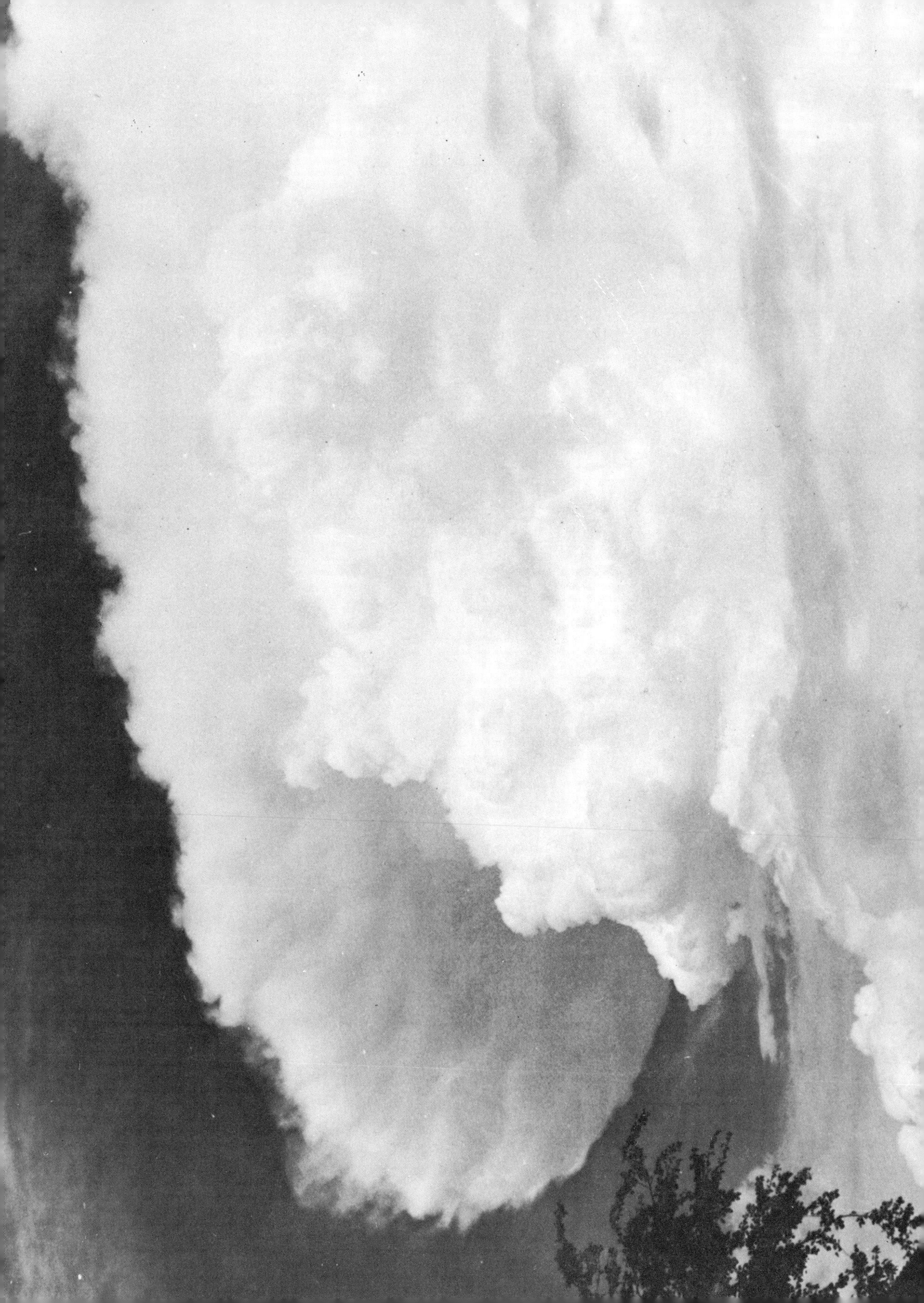

Condensation

As we have already seen, at a given temperature and pressure air can hold up to a given amount of water vapour, warm air being able to hold much more than cold (Fig. 18). When holding this maximum amount the air is said to be saturated and its relative humidity is 100 per cent. If saturated air is cooled, its vapour-holding capacity is reduced and in theory at least it must shed the excess vapour through condensation. But experiments have shown that condensation can only take place if there are tiny particles known as **condensation nuclei** in the air, and droplets then form around each of these nuclei. The concentration and chemical composition of the nuclei varies a great deal from place to place and from time to time. Over industrial areas there may be several million smoke particles per cubic centimetre, over rural areas there may be several thousand particles per cubic centimetre comprised mainly of dust from the weathering of soils, and over the sea there may be just a few hundred salt particles per cubic centimetre. The latter are very powerful nuclei, permitting condensation even before the air is saturated: they are known as **hygroscopic nuclei.**

Experiments show that the rate at which droplets grow upon their nuclei falls off quite sharply after reaching a radius of about 0.05 mm (0.002 inch), the size of most cloud droplets, but well below the radius of most drizzle (0.1 to 0.5 mm) (0.004 to 0.02 inch) and much less than the radius of the largest raindrops (2.5 mm) (0.1 inch). Raindrops and other forms of precipitation must therefore be formed by some process which causes these very tiny, buoyant cloud droplets to be drawn together to form larger, heavier units, capable of overcoming updraughts and reaching the ground without complete evaporation. The size of the cloud droplets is clearly relevant not only to these processes but to the shape of the clouds themselves. Tiny droplets will soon evaporate if they stray into the unsaturated air around a cloud and so the margins of cumulus clouds are often sharply defined since they are each surrounded by warm, unsaturated, sinking air. Cirrus clouds on the other hand are composed of ice crystals and because much lower concentrations of water vapour are required for saturation with respect to ice than with respect to water, ice crystals will not evaporate until low relative humidities are reached. This particular physical property is very important in the atmosphere. On the one hand it means that ice crystals will fall hundreds of feet before evaporating and can be scattered widely before they disappear; this causes the wispy, windswept look of cirrus clouds (Plate 13). Again, ice crystals will grow rather than evaporate when they fall from the base of a cloud, at least initially. Thirdly, if water droplets and ice crystals exist side by side, then the water droplets will tend to evaporate into the air which is unsaturated with respect to water but this same air may be saturated with respect to ice so that there will be sublimation (a change from vapour to solid) upon the ice crystals. The latter thereby grow at the expense of the former and shortly we shall see how important this property is in one of the processes of raindrop formation.

But before we discuss how raindrops are formed, there is one further 'change of state' that we must investigate a little more closely. This is the change from water to ice. Experiments show that water droplets in the atmosphere are very reluctant to freeze. Droplet clouds and fogs are commonly found at temperatures well below 0°C. They are then known as **supercooled** droplets. The reason for this delay in freezing is that as well as low temperatures, **freezing nuclei** are required and these nuclei appear to be much scarcer than condensation nuclei. Tiny splinters of ice will themselves act as very efficient nuclei so that once the process of freezing starts, it often spreads rapidly throughout the cloud although temperatures as low as minus 38°C are sometimes reached before freezing commences. Many minerals can act as freezing nuclei and some might be derived from the dust of meteoric storms beyond the earth.

13. High cirrus clouds

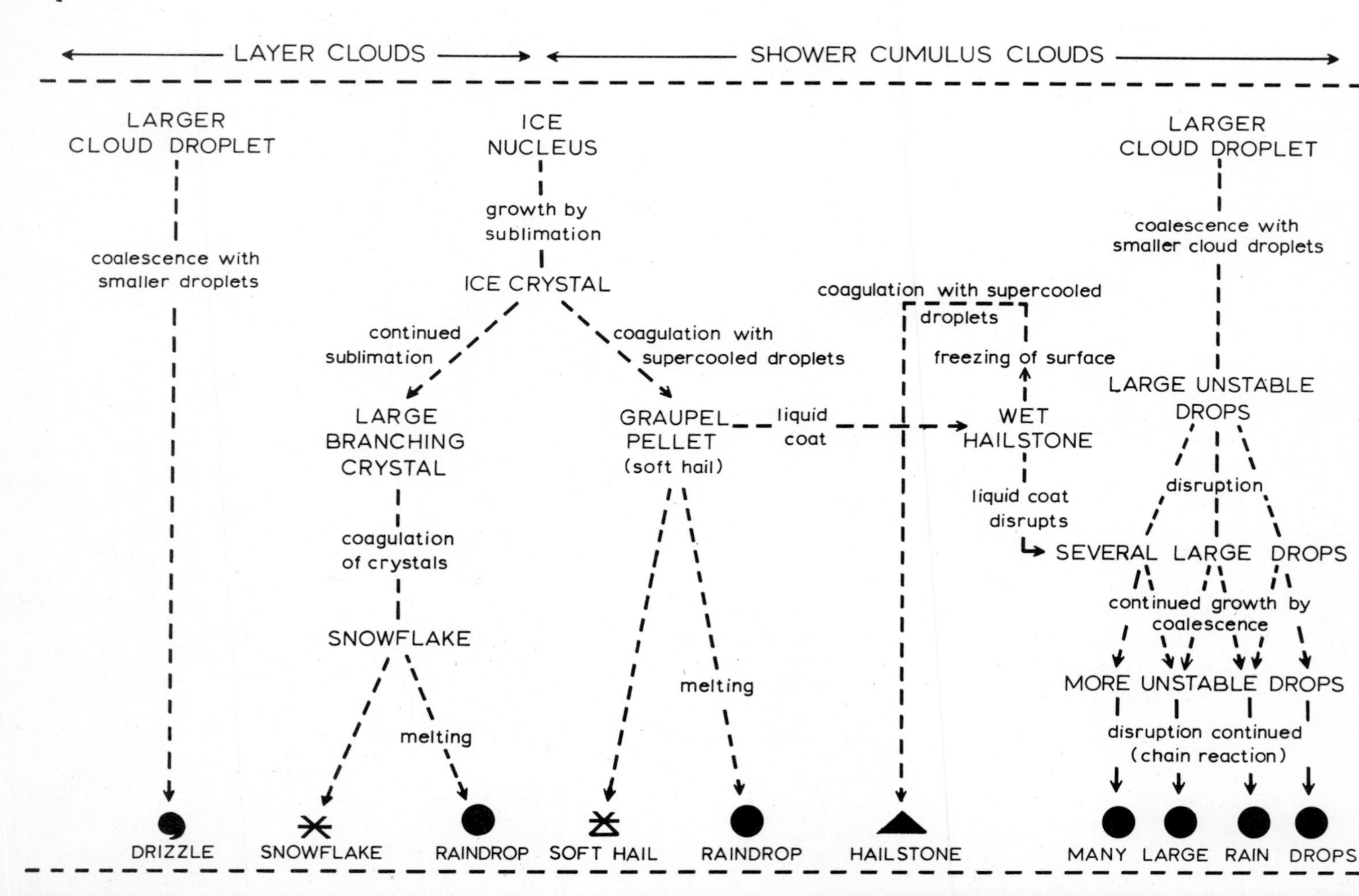

Fig. 22 Processes and sequences by which cloud droplets and ice crystals grow and combine to produce various forms of precipitation. After diagram in B. J. Mason, *The Physics of Clouds*, (Oxford University Press, 1957).

Rain

We have already seen that raindrops are formed not by condensation alone but by the combination of tiny cloud droplets. Before 1920, meteorologists were uncertain of the precise physical processes which drew together up to a million cloud droplets to form a single raindrop but now we know there are two main methods by which this is achieved and that these processes sometimes act separately and sometimes together. One of the methods, known as the **Bergeron-Findeisen process,** depends upon the instability of nearby water droplets and ice crystals. What happens in a deep cumulonimbus cloud is that the falling ice crystals grow by the condensation and freezing, or sublimation upon their surface, of vapour formed by the evaporation of supercooled water droplets. The ice crystals then combine into large snowflakes and below the freezing level they melt to form large raindrops (Fig. 22). This process is responsible for much of our heavy rainfall but it is obvious that it cannot explain all rainfall for it is a matter of common observation and particularly in the tropics, that rain frequently falls from clouds which contain no ice crystals. In these cases, the raindrops grow by the so-called **coalescence process.** This depends upon the sweeping up of tiny cloud droplets by a smaller number of larger droplets. Because of differences in frictional drag as they move up or down, droplets of different sizes move at different speeds and this leads to collisions, the large droplets growing at the expense of the smaller ones. Many of the unusually large cloud droplets are formed around particularly powerful condensation nuclei such as salt particles from sea spray. Also, most clouds are made up of a number of thermals or rising currents having different histories so that disparities in droplet sizes can also arise in this way. One has only to watch a cumulus cloud for ten minutes or so to see that some older parts are evaporating and disappearing whilst other parts are actively growing. The droplet size is often significantly different in areas having such contrasted histories.

Although the Bergeron-Findeisen and coalescence processes are quite dissimilar and although there is a tendency for the former to be mainly responsible for heavy rain from deep cumulonimbus clouds and the latter rather lighter rain from banks of stratus clouds, there is little doubt that the two processes often act in sequence, coalescence in the lower parts of clouds enlarging droplets produced by the melting of ice at higher levels.

One further and rather interesting method by which cloud or fog droplets are induced to precipitate is known as **fog-drip.** This happens when low cloud or fog drifts through trees and the droplets saturate the branches and leaves from which droplets then fall to the ground. In areas where fogs are seasonally persistent, such as in summer at low altitudes along the central Californian coastal belt, this process is very important. In this particular area, fog-drip accounts for about 10 inches (250 mm) of precipitation each year which is about half the annual total from all sources, and were it not for this source of moisture, the giant redwood forests of the coastal belt could not survive.

Hail

Hail is a major summer hazard in many parts of the world, particularly in mid-continental interiors in summer, and damage to crops amounts to hundreds of millions of pounds each year. This is not surprising because some of the largest stones are the size of cricket balls, weigh up to 5 lbs (2.5 kg) and hit the ground at over 100 m.p.h. (160 km/h). Sometimes, the hailstone is built up of an alternation of clear (high density) and white (low density) ice. These differences are produced in the intermediate levels of deep and active cumulonimbus clouds by the impact at different speeds of supercooled cloud droplets of different sizes. The larger droplets mould themselves around the hailstone before freezing, thus excluding most of the air so that a layer of clear ice is added to the hailstone. **Glazed frost** is a similar deposit on the earth's surface formed when rain falls upon frozen ground. Small droplets tend to freeze more quickly upon impact and a great deal of air is then trapped between

PLATE 14A

PLATE 14B

14 A (*above*) and B (*right*). These two photographs of rime frost were taken at the BBC's Holme Moss TV Transmitting Station on the Pennine ridge on 26 February, 1969, at a height above sea level of 1720 feet (522.0 m). A) shows a section of the 150 foot (46 m) aerial and B) a section of the support mast of the 750 foot (229 m) aerial. In both cases the ice 'spikes' are about 12 inches (305 mm) long and face in a NE direction.

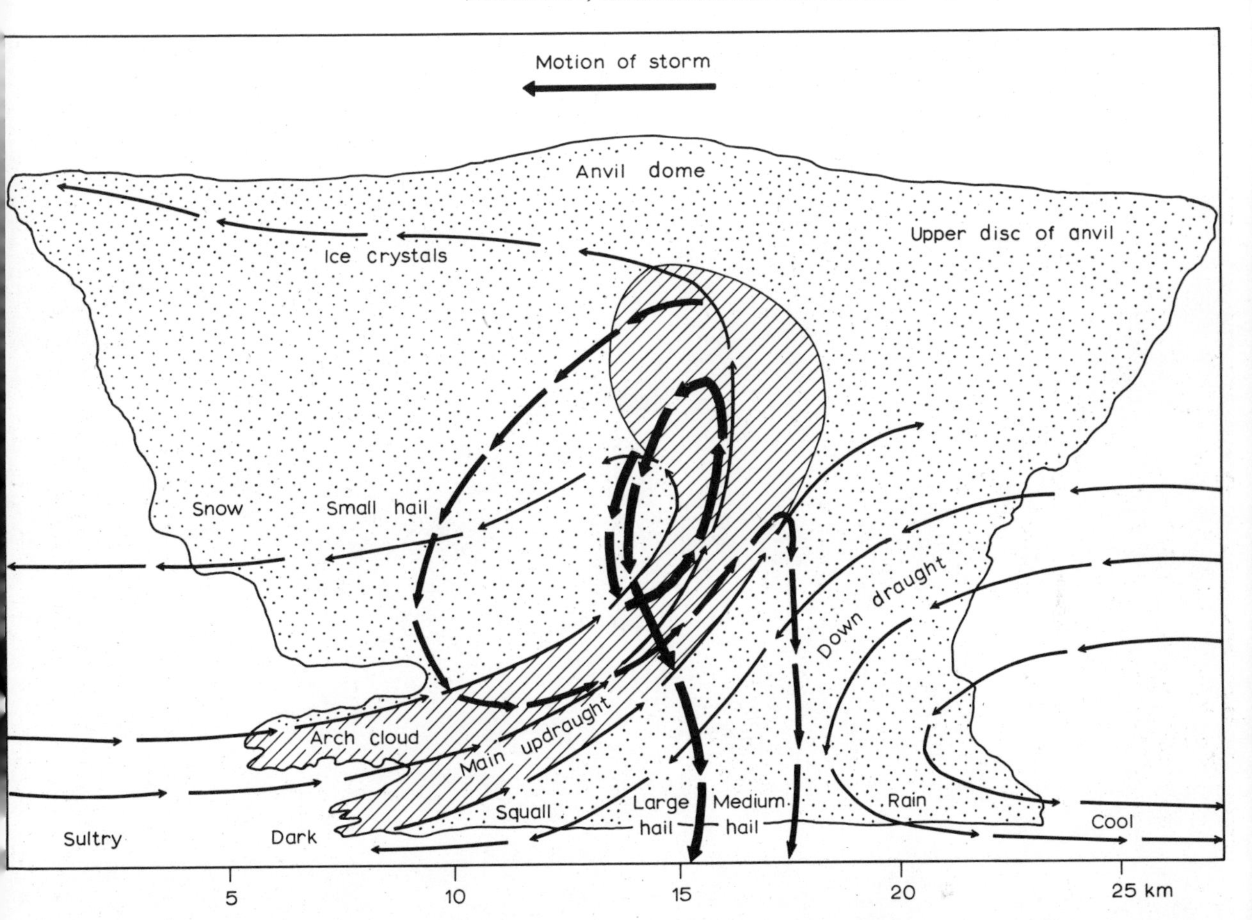

Fig. 23 Air movements in the direction of movement of a thunderstorm and the paths of hailstones growing inside the cloud. The largest stones (wide arrows) fall out of the main updraught (shaded) and then into it again so that they are carried up for a second time. Because of their size they rise only slowly in comparison with the supercooled droplets which, as a result, build up a thick layer of ice. After diagram in F. Ludlam, 'The Hailstorm', *Weather*, Vol. 16, 1961.

them so that the ice appears white and is of low density. This accumulation is very similar to the **rime frost** (Plate 14) which forms on the windward side of obstacles such as television masts and telegraph poles lying in the path of drifting supercooled fogs.

Ice is added to hailstones being carried in loop-like movements inside the cloud by violent currents (Fig. 23). If only one traverse is made, then the hailstones will be small and may melt entirely to give rain but if the stone gets carried up in a second ascent, it grows very quickly, since, being somewhat larger and heavier, it rises only slowly in the updraught and supercooled droplets collide with it in their millions.

Snow

Snowflakes consist of a loose cluster of individual ice crystals. These crystals can only form in relatively calm conditions inside a cloud and they are therefore associated with the less active cumulus and stratus clouds rather than the violent air movements inside cumulonimbus clouds. No two ice crystals are exactly alike although their general shape depends upon the temperature. At very low temperatures, small, needle-like crystals are produced while at higher temperatures, the familiar branching shapes are common and at temperatures only just below freezing, plate-like crystals are formed (Plate 15). The temperature of the air will also control the amount of available water vapour and the number of active freezing nuclei. At low temperatures there is a small amount of vapour competing for a large number of nuclei and the individual crystals are therefore small, while at higher temperatures there is a large amount of vapour available for only a small number of nuclei so that the crystals are large. Snow in upland areas tends to have a smaller average crystal size

than in lowland areas and because of this and the needle-like form, it drifts more easily in the stronger winds.

It has already been explained how snowflakes will continue to grow in the immediate subcloud layer but near the ground they will evaporate and perhaps melt. In evaporating, however, they will lose latent heat and this loss can sometimes keep their surface temperature below freezing though the air around is warm. **Sleet,** a mixture of snow and rain, will often reach the ground when air temperatures are as high as 3 or 4°C, that is in March and April in this country.

Other forms of precipitation

The term 'precipitation' covers not only rain, hail and snow but also moisture deposited from the air directly onto the earth's surface, that is, dew and hoar frost as well as the rime frost mentioned above. **Dew** is formed when the temperature of the earth's surface falls below the dew point, or the condensation temperature of the air: the water vapour of the air then condenses onto the cold surface. Because blades of grass have a large surface area and small mass they loose heat rapidly by night and their store of heat or thermal capacity is soon lowered. For this reason, dew tends to form most easily on meadows and lawns. In many parts of the semi-arid world, dew is a vital source of plant moisture. If the temperature of the ground falls below freezing either before or after condensation begins, then the deposit will be in the form of ice. A great deal of air will be trapped between the crystals and the white surface deposit is known as **hoar frost.**

15. Microscopic views of snowflakes, showing just a sample of the very wide variety of crystals produced by different meteorological circumstances.

Chapter 7

Weather types, depressions and anticyclones

Pressure and wind

The connection between pressure and weather was one of the first discoveries to be made following the invention of the barometer in the seventeenth century. The hall barometer or weather glass (Plate 2), still found in many homes, suggested a very simple relationship between these two variables, high pressure being associated with fine weather and low pressure with rain and strong winds. We now know that the relationships are by no means as simple and straight forward as this instrument suggests and that changes of pressure are generally far more informative than absolute pressure. In some respects it is wrong to regard pressure as anything more than one of the elements, as much a consequence of them as a cause. Even small-scale phenomena like thunderstorms can cause pressure changes, and the large scale wind circulations are of course responsible for the changes in large scale pressure patterns.

Surface pressures change in response to the net accumulations and removals and vertical motions of air at each level through the depth of the atmosphere. Areas of low pressure or depressions will deepen, that is, their central pressure will fall, if the ascent and divergence of air above is stronger than the convergence below; anticyclones will intensify if the convergence at high levels followed by descent more than compensates for the divergence near the earth.

Generally, winds blow parallel to the isobars with low pressure on their left in the northern hemisphere but there are other forces which may cause an angled trajectory towards or away from the high pressure. Only in very local circulations such as sea breezes does the air move directly from high to low pressure.

If one plots the path of a molecule of air over a period of about a week, one finds that it most frequently weaves its way round a series of alternating centres of high and low pressure, rather like a chain making its way around and between a series of cogs in a machine, some revolving clockwise and others anticlockwise. There is also a third dimension to these movements, the air sinking into the rear of a depression, then rising in a spiral anticlockwise motion to higher levels where it will be carried forward in one of the wavelike motions of middle altitudes before sinking again in a broad spiral-like motion of an anticyclone. In general, air tends to rise more quickly than it sinks and for this reason, areas of sinking air are more extensive than areas of rising air, that is, anticyclones tend to cover more of the earth's surface than do depressions. Also, for reasons connected with the balance of forces upon air moving in anticyclones, these pressure systems are individually larger than depressions (Fig. 30). There are no high pressure equivalents of intense low pressure systems such as tropical storms and tornadoes.

Weather Types

The air acquires its basic properties of heat, moisture and momentum mainly by exchanges with the earth's surface. It is not surprising therefore that where the physical properties of the earth's surface are fairly uniform and the motions of the air are slack, the air tends to acquire broadly similar properties, more particularly of the near-surface temperature, stability and humidity. In the period between World War I and II, great attention was paid to the recognition of these areas of general uniformity which were known as **air masses** but modern observation and analysis has brought a more cautious and realistic attitude. There are, in reality, no extensive atmospheric regions of near uniformity just as there are no geographically uniform regions of the earth. Everywhere there are gradients of the physical properties of varying intensity. The most intense gradients occur at **fronts,** the least intense in air masses or air streams.

Meteorologically, the air-mass concept is nowadays of only limited use in weather analysis and forecasting but the analysis of air-mass frequency and periodicity has proved a useful tool in climatology. Even here, however, the difficulty of distinguishing and recognizing air mass types has led to a more synoptic approach, the divisions being made in terms of a number of basic synoptic types and patterns and dominant wind directions. For Great Britain, Lamb has distinguished the following weather types; their frequency can be regarded as giving no more than a general indication of likely incidence, for as well as random changes from year to year, there are also long-period trends in their relative frequencies.

Weather types	*Percentage Annual Frequency, 1938–61*
Westerly and south-westerly	30.1
North-westerly	5.3
Northerly	8.6
Easterly	9.3
Southerly	11.9
Anticyclonic	15.3
Cyclonic	13.4
Unclassified	6.1

Depressions

Tornadoes

Areas of low pressure are known as depressions. They vary greatly in size from the tornado, a rotating column of unstable air only a few hundred yards across (Plate 16), to the mid-latitude frontal depression which may be between 1000 and 1500 miles (1500 to 2500 km) in diameter. Pressures at the centre of tornadoes are so low that buildings and other enclosed spaces literally explode as the twisting spiral of air moves over them. They occur at times of great instability in the lower atmosphere and are very common in the centre of continents in spring and summer. In America for instance, there are about 150 of the so-called 'twisters' each year and they can do an enormous amount of damage. **Waterspouts** over the seas and **dust-devils** over the deserts are all forms of the same phenomenon.

Tropical storms

Somewhat larger are the tropical storms (Fig. 24) which are found in all low-latitude ocean areas except the southern Atlantic where seas are cooler. They are mostly from 50 to 500 miles (80 to 800 km) across. The origin of these storms is not known for certain but they probably originate from waves in the tropical easterlies or trades when the atmosphere is unstable to very great heights. Subsequently they move around the western sides of the sub-tropical anticyclones over the oceans, deriving most of their energy from the conversion of latent heat to kinetic energy. For this reason tropical storms rarely survive long passages over the land where, in any case, surface friction causes the air to move more strongly across the isobars towards the low pressure centre of the storm which then fills. Even so, the shrieking winds and

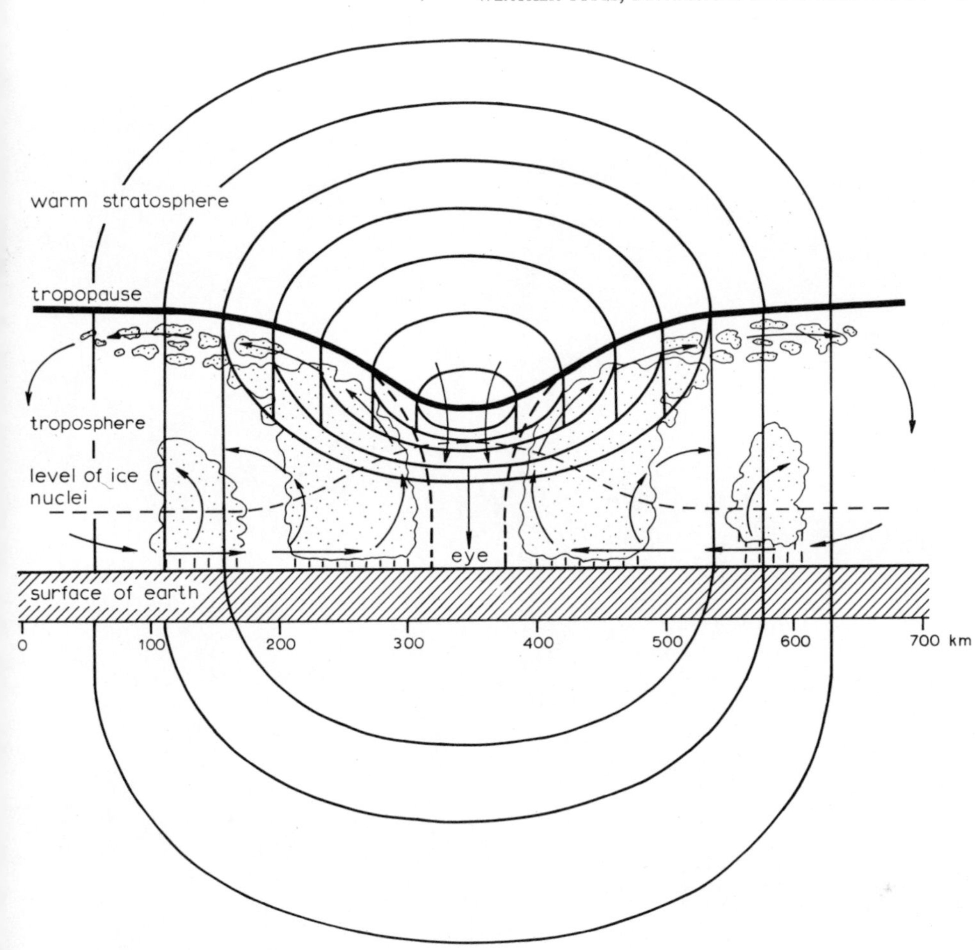

Fig. 24 Section through the low pressure vortex of a tropical storm. After diagram in R. S. Scorer, *Science Journal*, March 1966.

16. *(opposite)* A photograph of a tornado taken at Peshawar, India.

torrential rain can do enormous damage (Plate 17), made worse by the flooding of valleys and coastal areas by swollen rivers and huge sea waves driven before the storm. Characteristic of the tropical storm is the central 'eye', 10 to 30 miles (about 16 to 48 km) across, where skies are clear and the air is gently descending and warming, in complete contrast to the towering cumulonimbus clouds, violent winds and torrential rain around the eye. Most storms die out over the oceans on the northern side of the subtropical highs but occasionally they invade extratropical land areas where they may be regenerated by high level divergence. They may even develop fronts but their origins are betrayed by the characteristic circular form of the isobars.

Mid-latitude frontal depressions

When an airstream is embedded in a broad transverse thermal field, such as frequently occurs across the long waves in the westerlies (*see* Chapter 5), the atmosphere in the zone is said to be **baroclinic.** In the northern hemisphere, the temperature will most commonly fall from right to left across the airstream and the windspeed will then increase with height to a maximum in the jetstream centred near the tropopause. Such an airstream is **baroclinically unstable,** and forces owing to the temperature gradient across the wind will cause it to meander. As long as the waves are small (generally less than 300 miles (500 km) long) they will tend to remain so, although they will move forward rather quickly,

江南肉食公司
KEEN ON

perhaps 600 miles (1000 km) in a day. They are known as **stable waves** and frequently cause a short period of bad weather. With wavelengths between about 300 and 1000 miles (500 and 1600 km), (depending upon the speed of the wind, the temperature gradient and the breadth of the current) the waves amplify and are known as **unstable waves.** The rapid changes in curvature cause the air to decelerate when moving into the cyclonic curvature of a trough and accelerate when moving out of the trough into the anticyclonic curvature of the following ridge. This results in convergence behind the trough and divergence ahead (Fig. 15). The motions will normally be linked through sinking and rising air respectively to contrary movements nearer the earth, that is, divergence behind the trough and convergence ahead. The column of air linking the two levels will shrink behind the trough line and deepen ahead (Fig. 25).

Unstable waves are thus associated with very fundamental systems of air movement which basically account for the anticyclone we normally find upstream of a trough (that is downstream of a ridge) and the depression found downstream of a trough (that is upstream of a ridge) in the wave pattern (Figs. 15 and 26). Pressure falls ahead of the trough and increases behind because the lighter winds nearer the earth than higher up allow only imperfect compensation between the two levels. In the area of convergence near the ground, there is a concentration of the spin which the air has because of the earth's rotation and so a depression is developed with air revolving in the same way as the earth, that is, anticlockwise round its centre in the northern hemisphere. For the opposite reasons, anticyclones with clockwise winds are found behind a trough (Fig. 25).

When a wave is amplified, then the strong convergence behind the trough line in the middle and upper troposphere concentrates the isotherms between the north-easterly flow in the rear of the trough and the south-westerly flow ahead. The jet stream is further strengthened by this concentration of the isotherms (Fig. 26). Nearer the earth the convergence ahead of the trough will concentrate the isotherms here and so a sloping frontal zone will be created. We now see that fronts are more realistically regarded as secondary consequences of mid-latitude depressions than as necessary components of their formation as previously regarded. One further point is that clouds and precipitation, though frequently resulting from the upglide of air along the frontal surface, are clearly not always formed in this way, for at higher levels belts of cloud are frequently separated by clear air from the frontal surface. It seems likely that this cloud lies in areas of rising air and high level divergence on the warm air side of the zone of convergence which is the line of the frontal surface. The release of latent heat as water vapour condenses to form clouds can account for up to half the heat energy of a depression, thus encouraging further uplift.

At the ground, the cold and warm fronts mark leading edges of the cold air behind and the warm air ahead of the trough respectively (Fig. 26). Between them lies the warm sector whose area (at the surface) is gradually diminished by the faster forward movement of the cold front. Beginning at the centre, the cold front catches up with the warm front and according to whether the air behind the cold front is colder or warmer than the air ahead of the warm front, it either cuts under or rises over the warm front. The warm air is lifted above the ground (Fig. 27) and the zone of separation between the cold air behind the cold front and the cold air ahead of the warm front is known as an **occlusion** which is also marked by vertical motions and hence cloud and sometimes rain.

17. Tornado damage in Hong Kong.

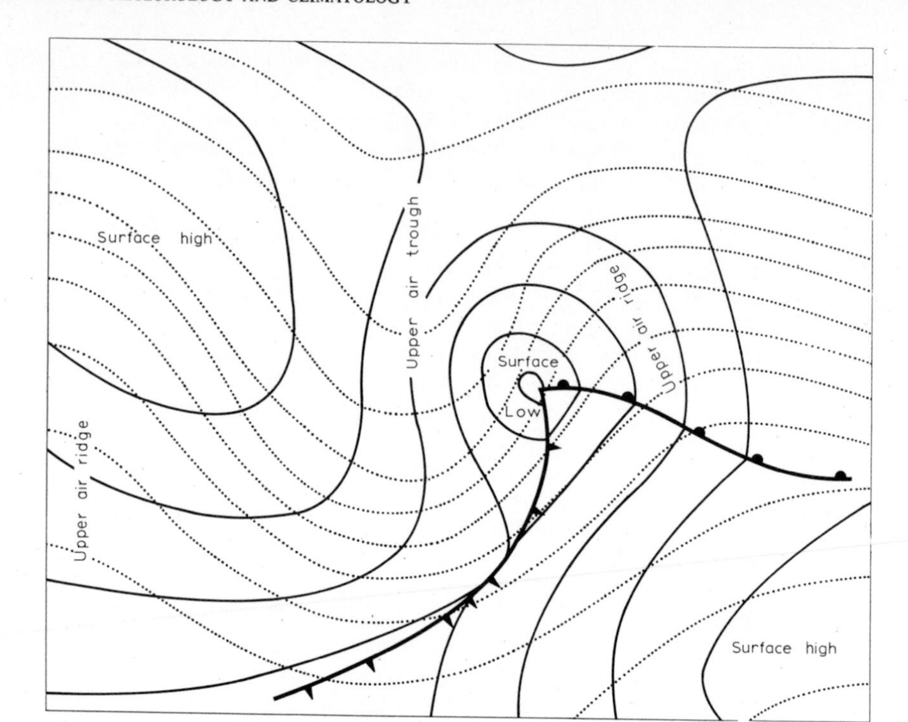

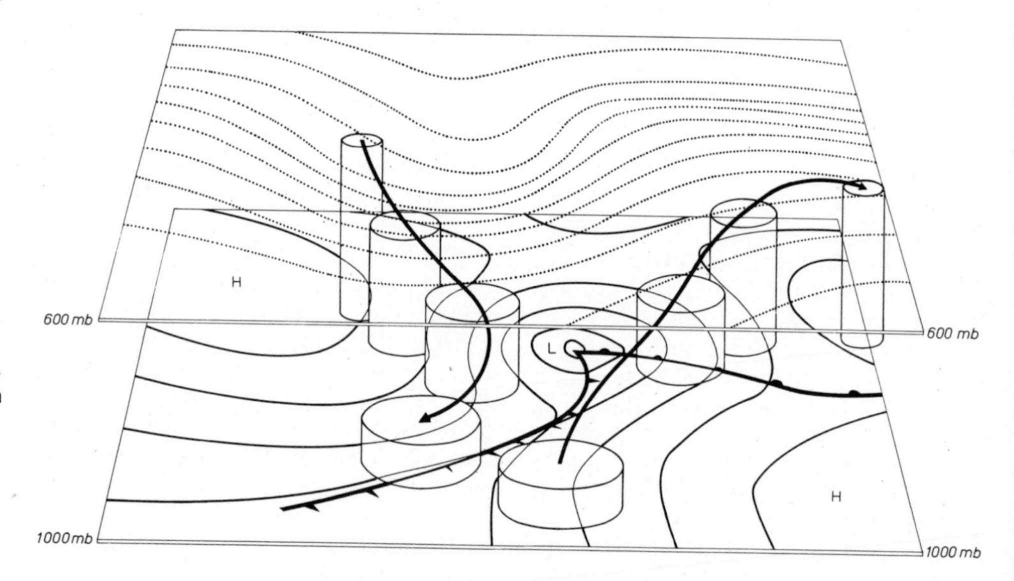

Fig. 25 The relationships between the pressure patterns and circulations in the upper westerlies (600 mb) and those near the surface (1000 mb). In the upper diagram, dotted lines represent a trough in the upper westerlies. Ahead of the trough, near the ground, there is a depression (shown by continuous lines representing isobars, and by cold and warm fronts) and behind the trough line there is a near-surface anticyclone.

In the lower diagram the 600 mb pressure surface is shown above the 1000 mb surface. Behind the 600 mb trough line, cold air sinks, diverges and curves anticyclonically; ahead of the trough line, between the cold and warm fronts, in the warm sector of the depression and continued ahead of the warm front, the air columns are stretched vertically and their (cyclonic) vorticity is increased before decreasing again in the western limb of the next wave. After diagram in E. Palmén and C. W. Newton, *Atmospheric Circulation Systems* (Academic Press, 1969).

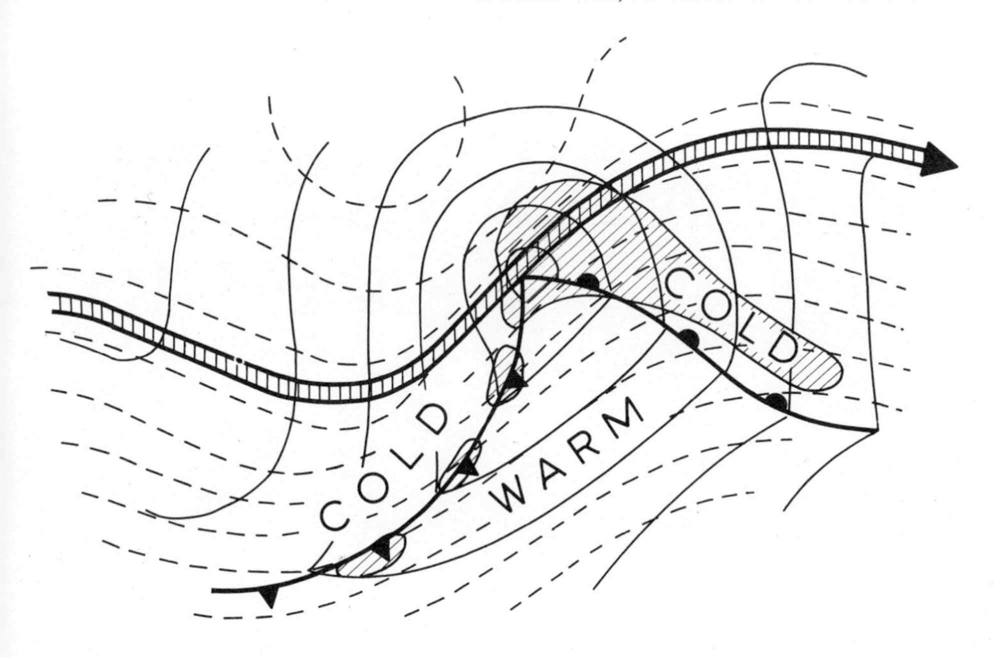

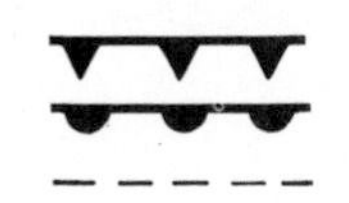

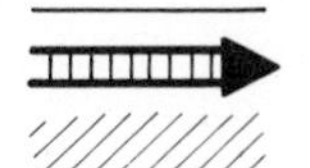

Fig. 26 Diagram of a mature, frontal, mid-latitude depression in relation to a trough in the upper westerlies and its associated jetstream.

Fig. 27 *(below)* Sections through a cold and warm occlusion. After diagram in D. Pedgeley, *A Course of Elementary Meteorology* (H.M.S.O., 1962).

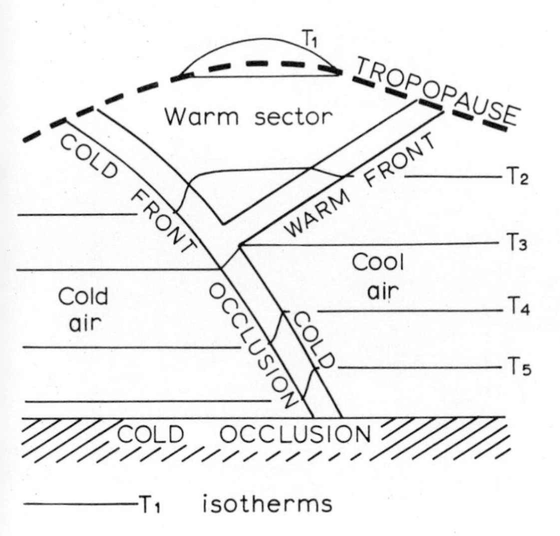

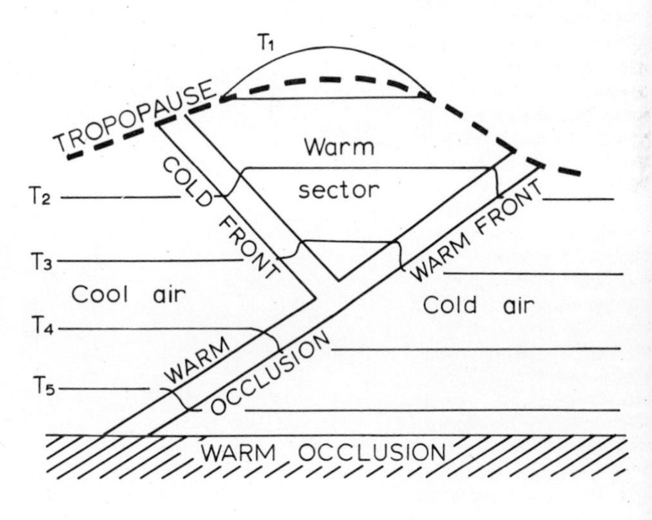

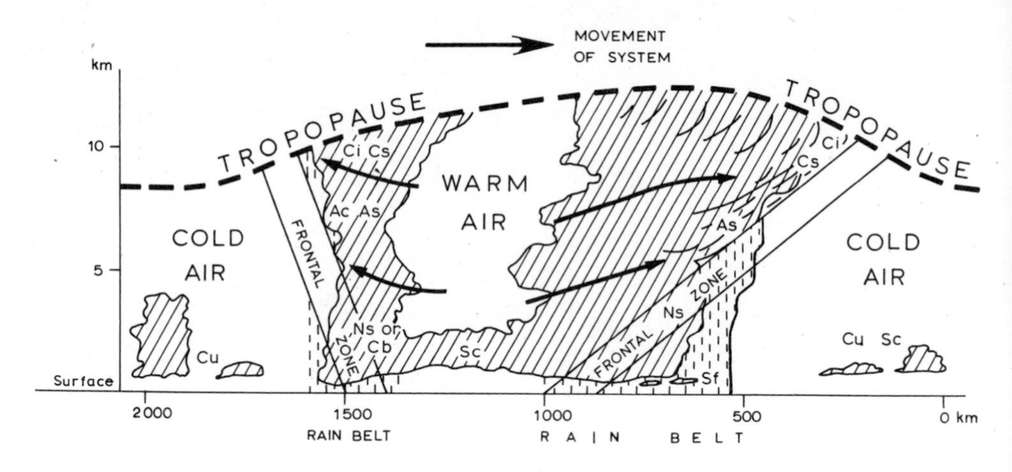

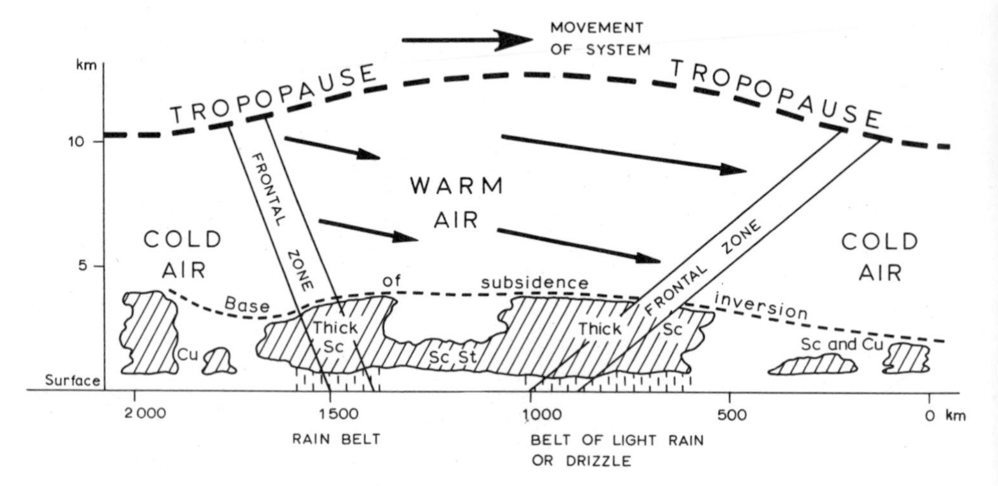

Ci — Cirrus
Cs — Cirrostratus
As — Altostratus
Ac — Altocumulus
Ns — Nimbostratus
St — Stratus
Sc — Stratocumulus
Sf — Fractostratus
Cu — Cumulus
Cb — Cumulonimbus

Fig. 28 Sections through two depressions. In both, frontal zones separate the cold and warm air; the cold front lies behind and the warm front ahead of the warm air. The wind arrows show air movement relative to the fronts. In the upper diagram (*a*), there is general air ascent throughout the depression and there are deep clouds, especially close to the fronts, known as *ana-fronts*. In the lower diagram (*b*), there is widespread subsidence producing a low inversion which acts as a lid to rising currents and cloud. These fronts are known as *kata-fronts*.

Recent observations on fronts have shown them to be far more complicated than the simple so-called 'Norwegian' model suggested. The idea of a broad, uniform belt of upgliding air with rain at the warm front is far too simple for instance, for there is clearly a fairly complicated cellular structure of rising and sinking currents which are responsible for the patchiness and bands of the rainfall which detailed analyses of precipitation have shown us to be typical. Also, the nature of frontal weather clearly depends upon the overall character of vertical movements in their vicinity and meteorologists have made a broad distinction between **ana-fronts** where the air is generally rising and **kata-fronts** where it is broadly descending (Fig. 28). In the first case, the sinking air will be warmed by compression and this makes the atmosphere rather stable, whilst in the second situation, the uplift will favour instability. The differences will materially affect the weather. In most cases, fronts tend to be ana-cold fronts near the centre of the depression but kata-cold fronts further away.

The depression itself usually follows a curving path, curving cyclonically (in an anticlockwise direction) in the northern hemisphere. The actual path is very close to the direction of the mean isotherms for the layer between the ground and about 500 mb (which is half-way through the mass of the atmosphere) at about 18 000 ft (5500 m). The precise movement of the depression is, in fact, strongly influenced by the position of deep, slow-moving anticyclones embedded in the atmosphere, rather like whirlpools move around areas of calmer water in a stream. These so-called **blocking highs** are frequently formed when the waves in the westerlies become so exaggerated that the crestal areas of the ridges break away and then stagnate, after when they tend to slow down advancing depressions and steer them around their margins. In a similar way, the apices of troughs are often separated and drift south-eastwards; in the north Atlantic region they enter the Mediterranean as pools of cold air which may, near the ground, acquire fronts.

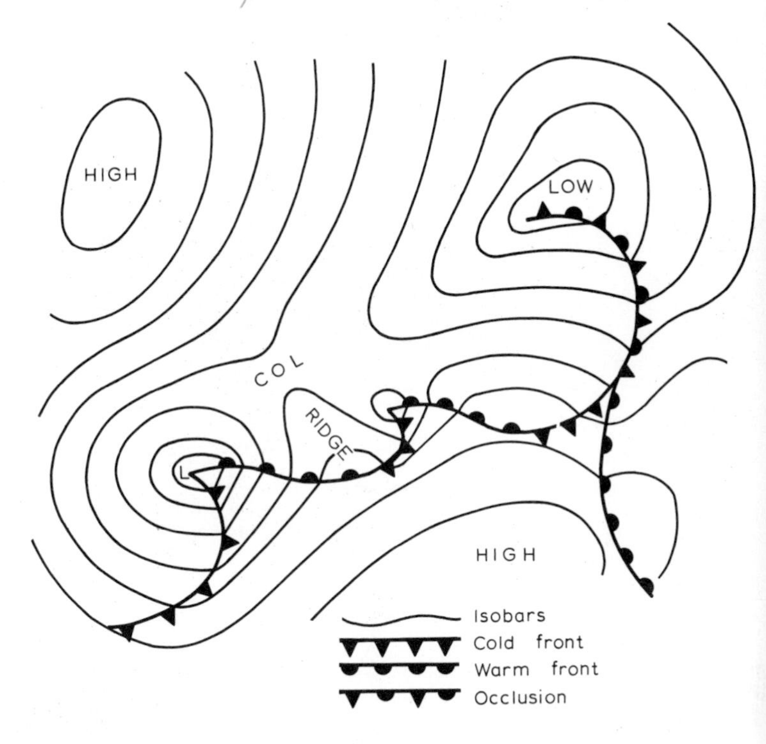

Fig. 29 A family of depressions.

Though they can occur singly, frontal depressions are frequently linked together as a series of waves which increase in maturity from west to east (Fig. 29). The alignment of this so-called **family of depressions** is commonly from south-west to north-east and Fig. 9 shows that it is frequently related to the thermal gradients produced between the cold air of the western limb and the warm air of the eastern limb of a tropospheric long wave. These temperature gradients are intensified by the thermal contrasts between land and sea and by ocean currents, all of which help to control the mean position of the main frontal zones (Fig. 9).

Non-frontal depressions

Although many of the depressions in middle and high latitudes form in the manner described above, many others do not and are non-frontal. **Thermal lows** are such a group. They are formed by the pressure changes induced by local heating and vary in size from the small, shallow depressions which form over East Anglia, to the intense depressions which form over Iberia and northern India in summer.

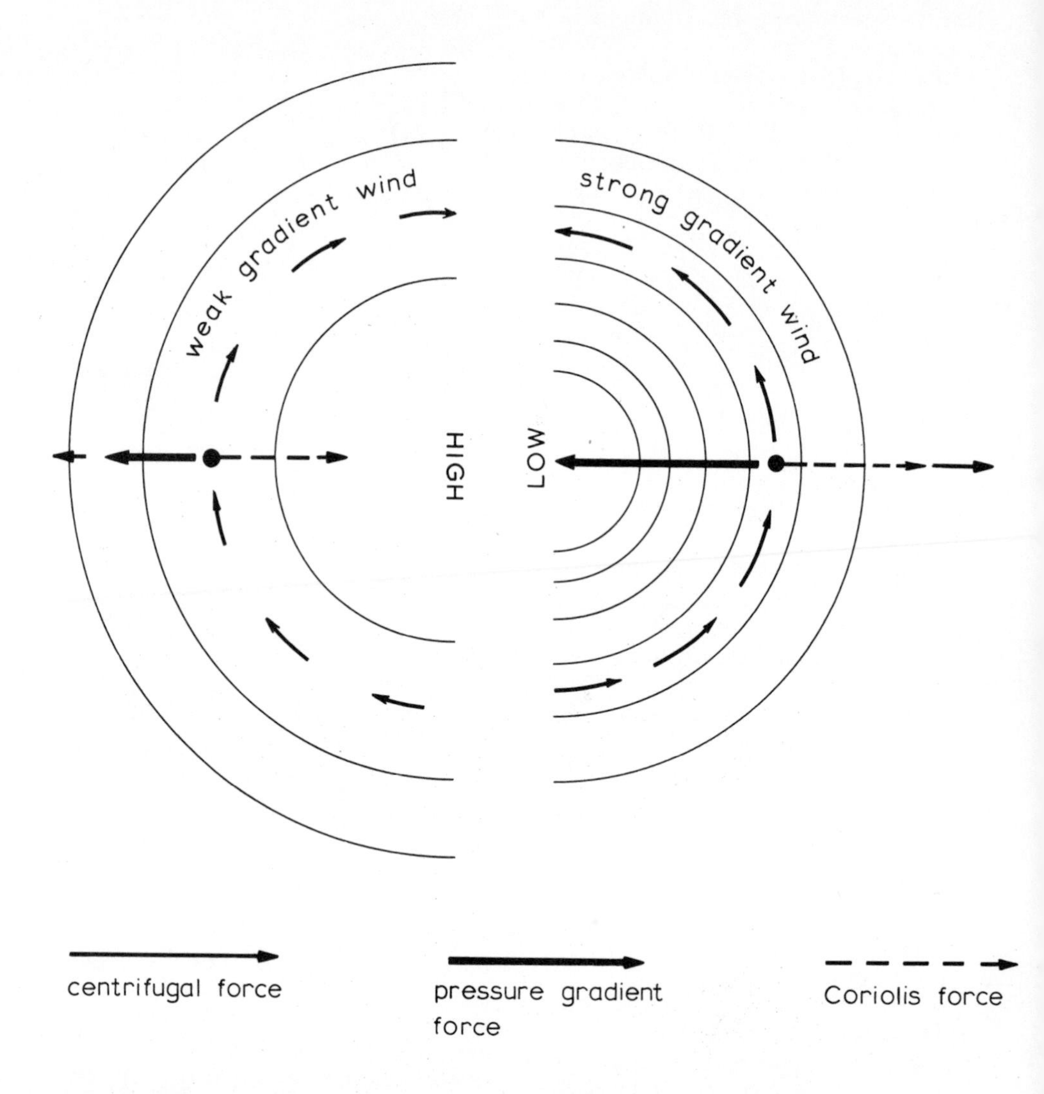

Fig. 30 The balance of forces on winds rotating around centres of high and low pressure. Because of these balances, anticyclones are generally larger and their winds lighter than in depressions. The air in anticyclones also revolves in the opposite sense.

Polar air depressions are often a particular type of thermal low, formed by the surface heating of an unstable polar airstream in the rear of a mature or occluding depression. The centre of low pressure, known as a **secondary,** will initially revolve around the major depression or **primary.** Being shallow, it will be fast moving and being associated with in-blowing and rising air it will give unsettled weather and probably heavy rain. Forming and moving so rapidly, these secondary depressions pose something of a problem to weather forecasters.

Lee depressions or **orographic lows** are another type of non-frontal depression although like others, they sometimes bring together air of different temperatures, thereby creating fronts. These depressions are formed by the movement of air over a topographic barrier leading to vertical contraction, divergence and hence reduced spin over the crest of the barrier and vertical expansion, convergence and increased spin to the lee (Fig. 10). They can be compared to the eddies and whirlpools which form downstream of a protruding rock in the bed of a river and are commonly found to the lee of the world's major mountain ranges although contrasts of airmasses across these mountains cause many of them to become frontal.

Anticyclones

The atmospheric pressure in anticyclones increases towards their centre and in the northern hemisphere, winds rotate in a clockwise direction. These winds are controlled by a balance of forces: the pressure gradient term and the apparent force produced by rotation (known as the centrifugal force) are both directed outwards from the centre and oppose the force owing to the earth's rotation known as the Coriolis Force (Fig. 30). This force balance is such that it can only exist if the centrifugal and pressure gradient terms are small, that is, if the curvature of the isobars is gentle and they are widely spaced. For these reasons, anticyclones are generally large and their winds are light.

The daily weather charts show that a useful distinction can be made between those middle latitude anticyclones which are interspersed between and move with the depressions and the generally larger and much more stationary anticyclones which are found, for instance, in the subtropics. The former are formed, as already explained, by high-level air convergence and subsidence on the downstream limb of a ridge in the tropospheric westerlies while the latter form in the zone of intense subsidence (and thereby stability) on the poleward side of the Hadley cell.

Anticyclones can also be divided into cold-centred and warm-centred types. Because pressure falls rapidly with height in cold air, cold-centred anticyclones tend to be rather shallow, and often swiftly moving, while warm-centred anticyclones, for the opposite reasons, tend to be deep and fairly stationary.

Anticyclones are traditionally associated with fine, sunny weather but observation shows the relationship to be far from certain and a far better correlation than this is that between fine, sunny weather and anticyclones for, in winter especially, high pressure can often result in periods of fog and low cloud, probably with drizzle. In both cases, it is the very gentle subsidence of air in the anticyclone which is the main control upon the weather. In the tropics it produces the trade wind inversion at heights of from 6000 to 10 000 ft (1800 to 3000 m) which inhibits the growth of clouds and is mainly responsible for the aridity of these areas. In middle latitudes the subsidence near the centre of anticyclones can lead to clear skies and strong sunshine but equally, if temperatures are low and humidities high, condensation can occur here to give persistent fog or low cloud beneath the subsidence (and perhaps radiation) inversion. The stability of the inversion can also prevent the escape of pollution from domestic and factory chimneys, and this may give rise to urban smog.

Chapter 8

Patterns of climate

Figures 13 and 14 show the broad global distribution of pressure and winds at the earth's surface and Fig. 31 the seasonal and annual distribution of precipitation. The distributions are complicated by the earth's varied topography, but behind the local complications there is clearly a repeated pattern to the world's climates. This pattern reflects the fundamental control of the general circulation of the atmosphere discussed in Chapters 4 and 5.

The world's climates have been classified in a number of ways, one of the best known of which was devised by W. Köppen in the early twentieth century. Fig. 32 shows the distribution of the various climatic types according to this classification and the repetition of broadly similar climates in comparable latitudes and continental positions in striking. Nevertheless, more detailed inspection shows that there are interesting differences between the continents in the detail of their climates, more particularly in low latitudes and for this reason they will be dealt with in more detail.

Low latitude climates

Low latitude climates have received a great deal of attention in recent years and the previous notions of their simplicity and uniformity have been largely exploded. For instance, so-called **'easterly waves'** are common during summer and autumn in the trade-wind areas of the two hemispheres. The waves appear on the pressure charts as troughs of low pressure extending polewards from the equatorial low-pressure zone (Fig. 33). As the wave moves westwards through the trades there is near-surface divergence and subsidence ahead of the trough line giving generally fine weather with scattered cumulus cloud but rather poor visibility. Behind the trough line there is convergence and uplift of moist air giving thundery showers from cumulonimbus clouds and improved visibility. The 'tornadoes' of West Africa are local examples of this type of phenomena. In winter and spring the trade wind inversion is lower and this inhibits the growth and activity of these waves.

Again, the **intertropical convergence zone** (ITCZ) has been the subject of careful study and, as so often happens, it has proved to be a more complex phenomenon than was originally supposed. Only over the Pacific ocean is the convergence of the trades in the two hemispheres very intense. Elsewhere, especially from July to September, the trades are separated by a belt of calms known as the Doldrums or, in the summer hemisphere over the land and over the Indian Ocean, by a shallow belt of westerly winds lying between the Equator and the thermal equator (Fig. 8). These are not the south-east trade winds deflected on crossing the Equator as has sometimes been supposed. The maximum uplift in the Hadley cells is often separated from the 'thermal equator' and its belt of thermal lows known as the **equatorial trough.**

The poleward sections of the equatorial westerlies are known as the 'Indian monsoon' which has also been proved to be a more complex phenomenon than was once supposed.

Fig. 31 The seasonal and annual distribution of global precipitation. A *(opposite)* shows summer distribution, B (p. 60) shows winter distribution, and C (p. 61) shows annual distribution.

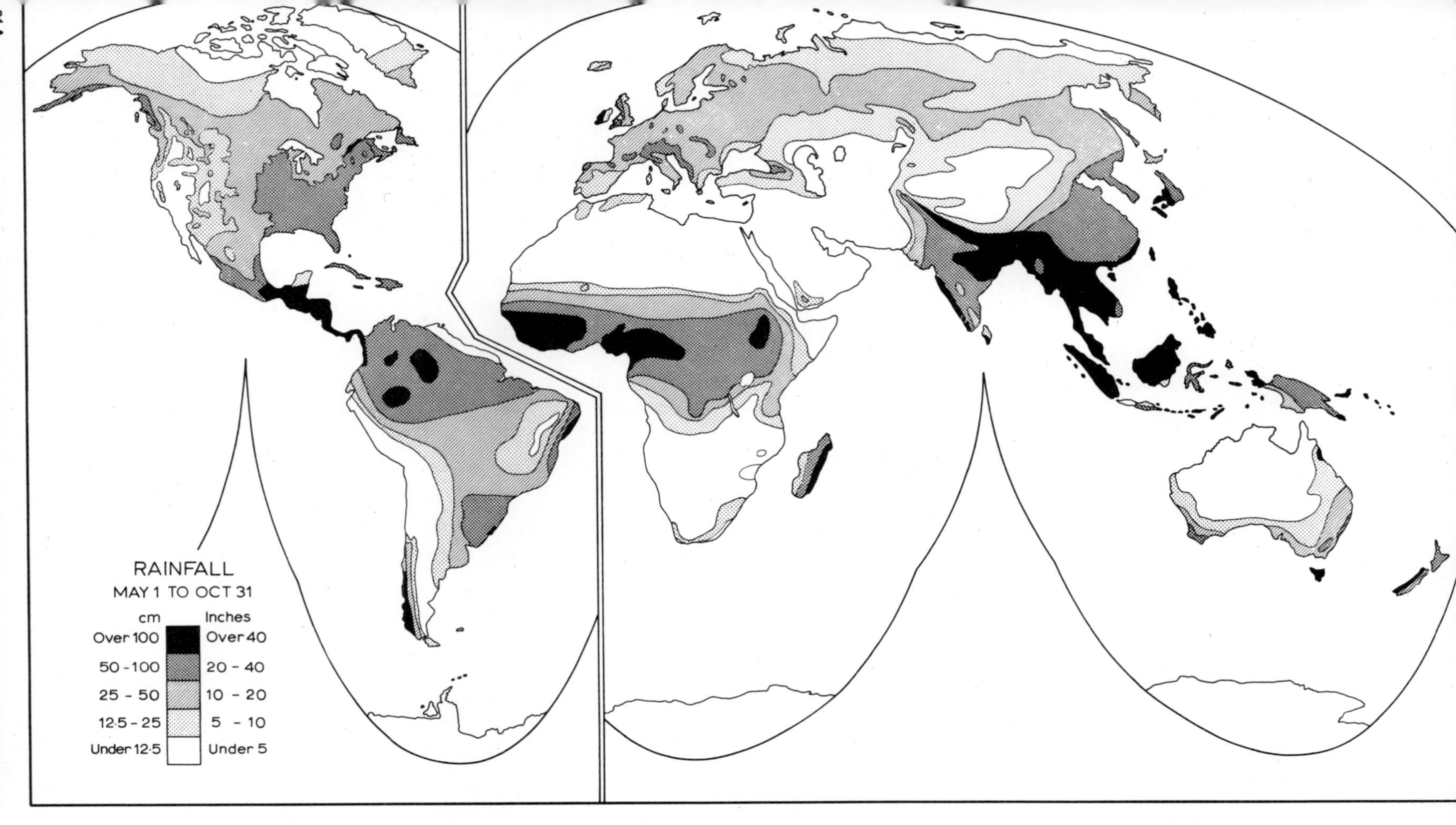
RAINFALL
MAY 1 TO OCT 31
cm
Inches
Over 100
Over 40
50 - 100
20 - 40
25 - 50
10 - 20
12·5 - 25
5 - 10
Under 12·5
Under 5

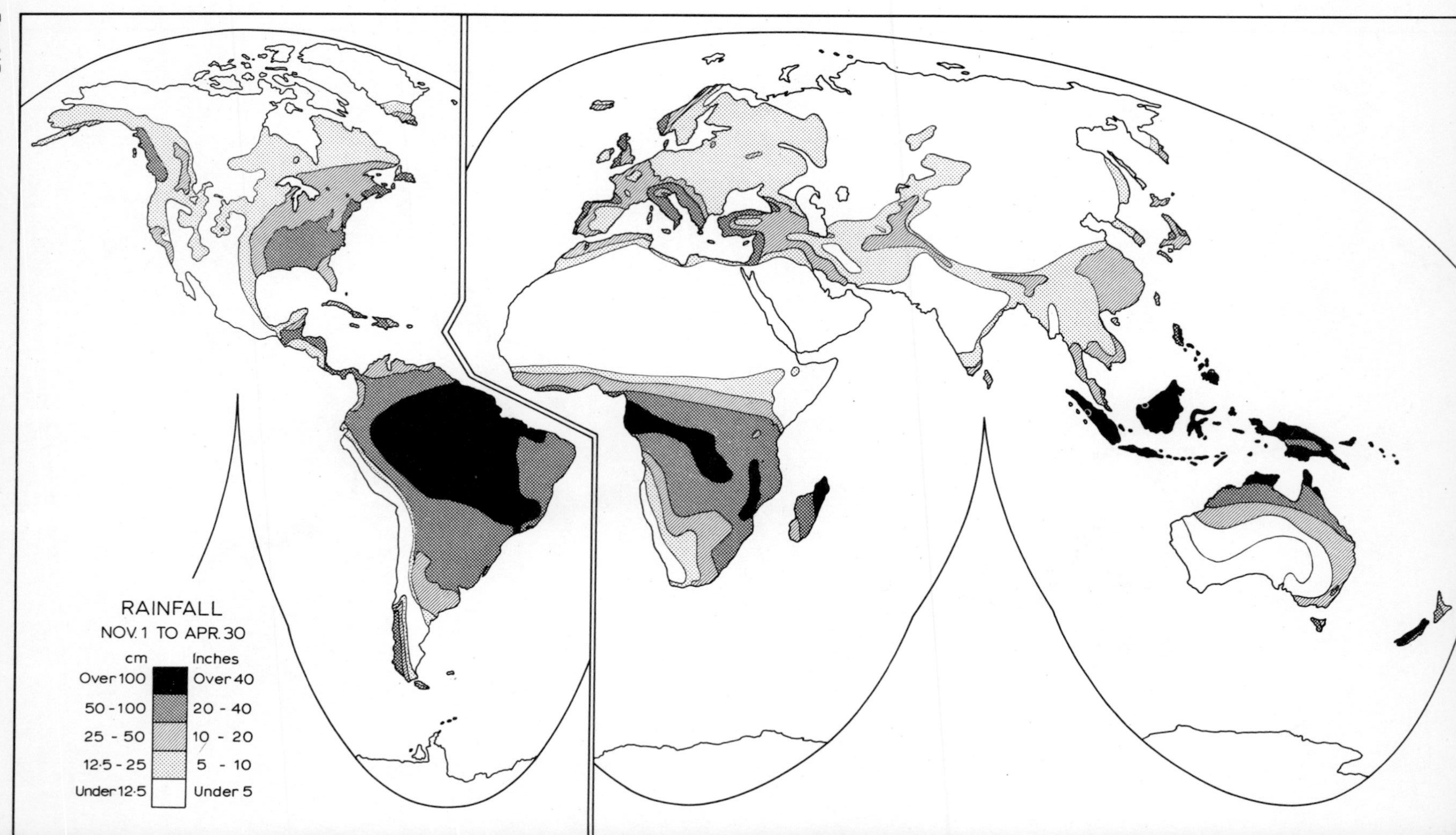
RAINFALL
NOV. 1 TO APR. 30
cm
Inches
Over 100
Over 40
50 - 100
20 - 40
25 - 50
10 - 20
12·5 - 25
5 - 10
Under 12·5
Under 5

Fig. 31B

Fig. 31C

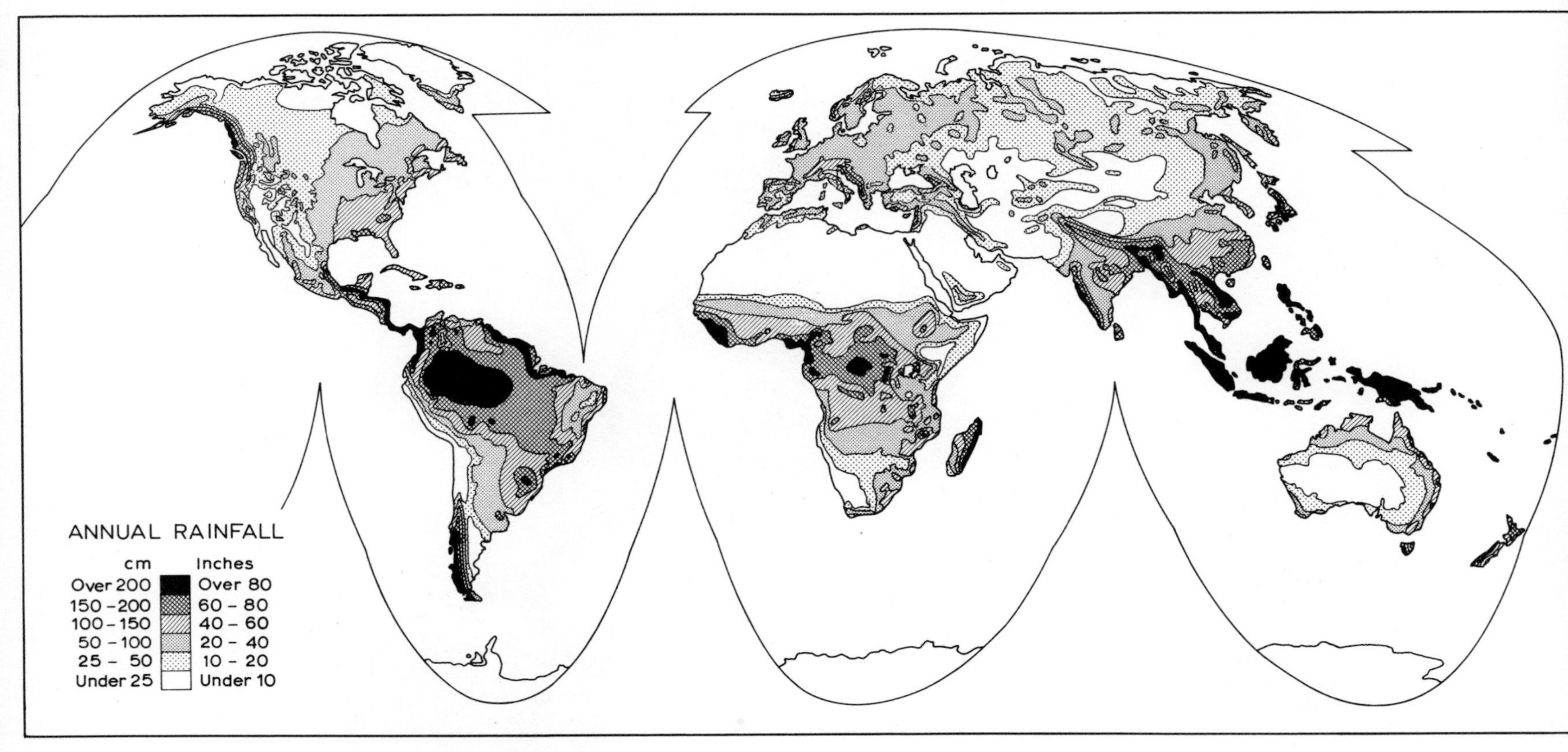
ANNUAL RAINFALL
cm
Inches
Over 200
Over 80
150 – 200
60 – 80
100 – 150
40 – 60
50 – 100
20 – 40
25 – 50
10 – 20
Under 25
Under 10

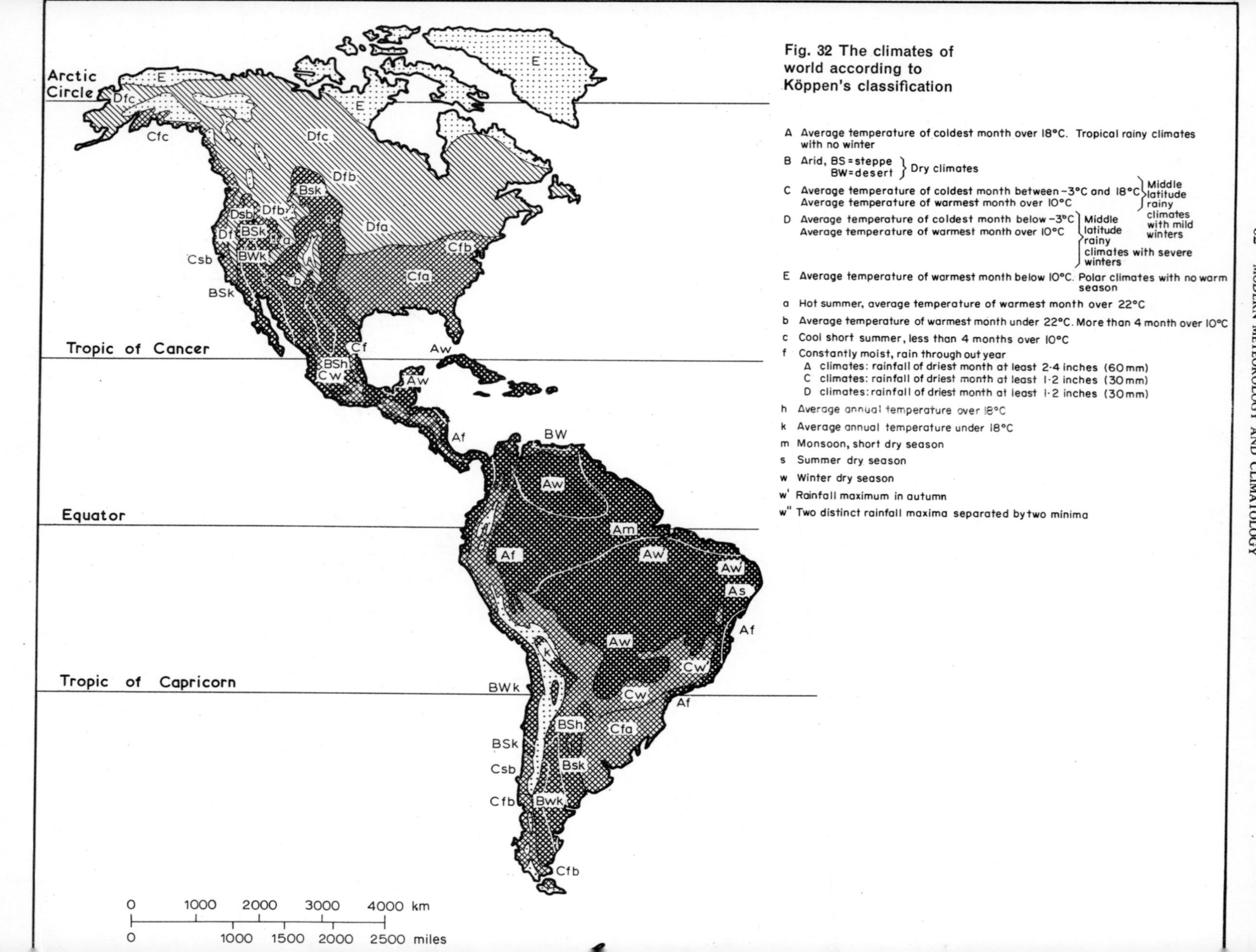

Fig. 32 The climates of world according to Köppen's classification

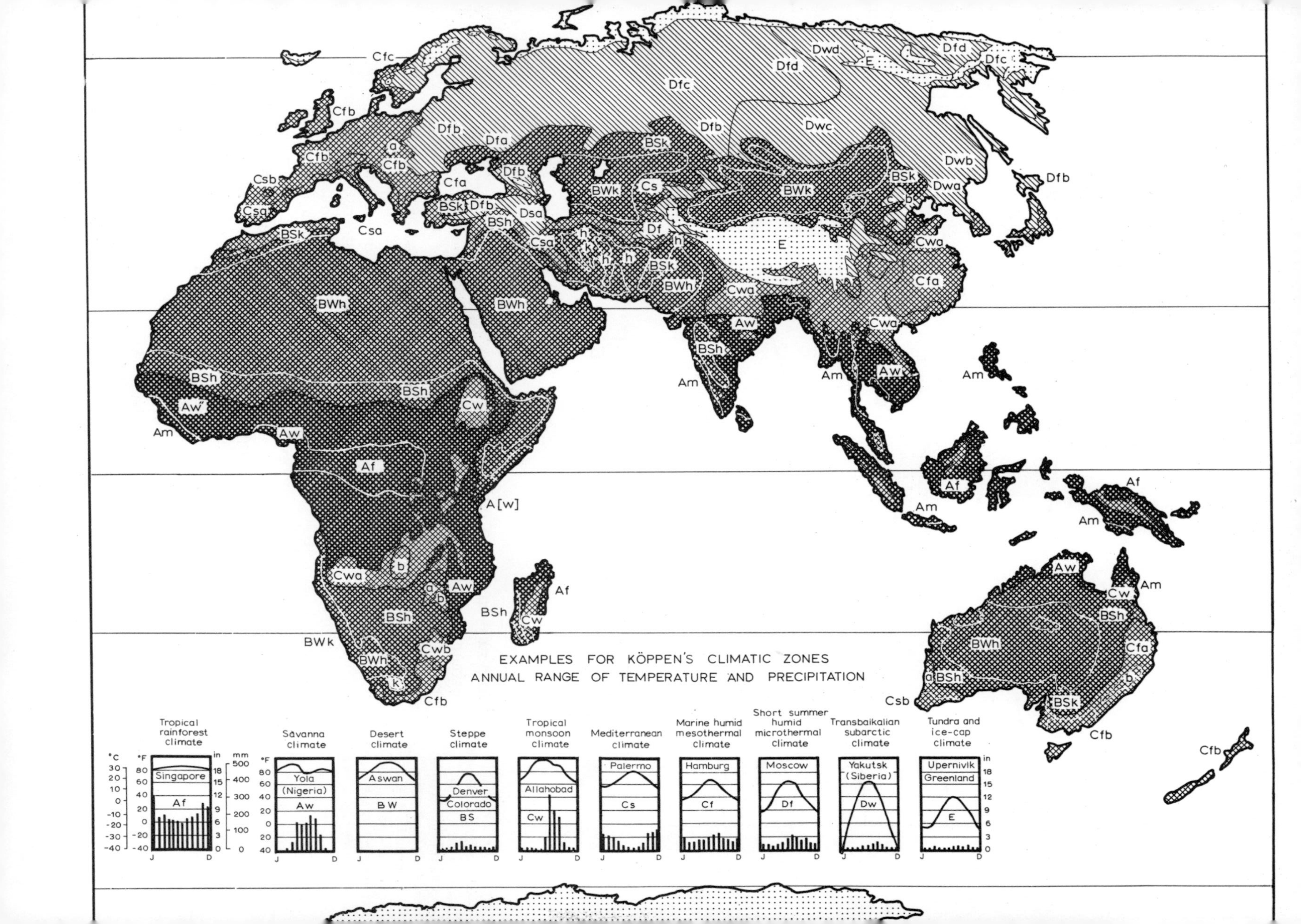
EXAMPLES FOR KÖPPEN'S CLIMATIC ZONES
ANNUAL RANGE OF TEMPERATURE AND PRECIPITATION
Tropical rainforest climate
Singapore
Af
Savanna climate
Yola (Nigeria)
Aw
Desert climate
Aswan
BW
Steppe climate
Denver Colorado
BS
Tropical monsoon climate
Allahabad
Cw
Mediterranean climate
Palermo
Cs
Marine humid mesothermal climate
Hamburg
Cf
Short summer humid microthermal climate
Moscow
Df
Transbaikalian subarctic climate
Yakutsk (Siberia)
Dw
Tundra and ice-cap climate
Upernivik Greenland
E
°C
°F
in
mm
Cfc
Dfd
Dwd
Dfc
Dfb
Dwc
Dfa
Cfb
Csb
Csa
Cfa
BSk
BWk
Cs
Dsa
BSh
Df
BWh
Cwa
Dwb
Dwa
Aw
Am
Aw″
Af
A[w]
Cw
BWk
Cwb
Csb

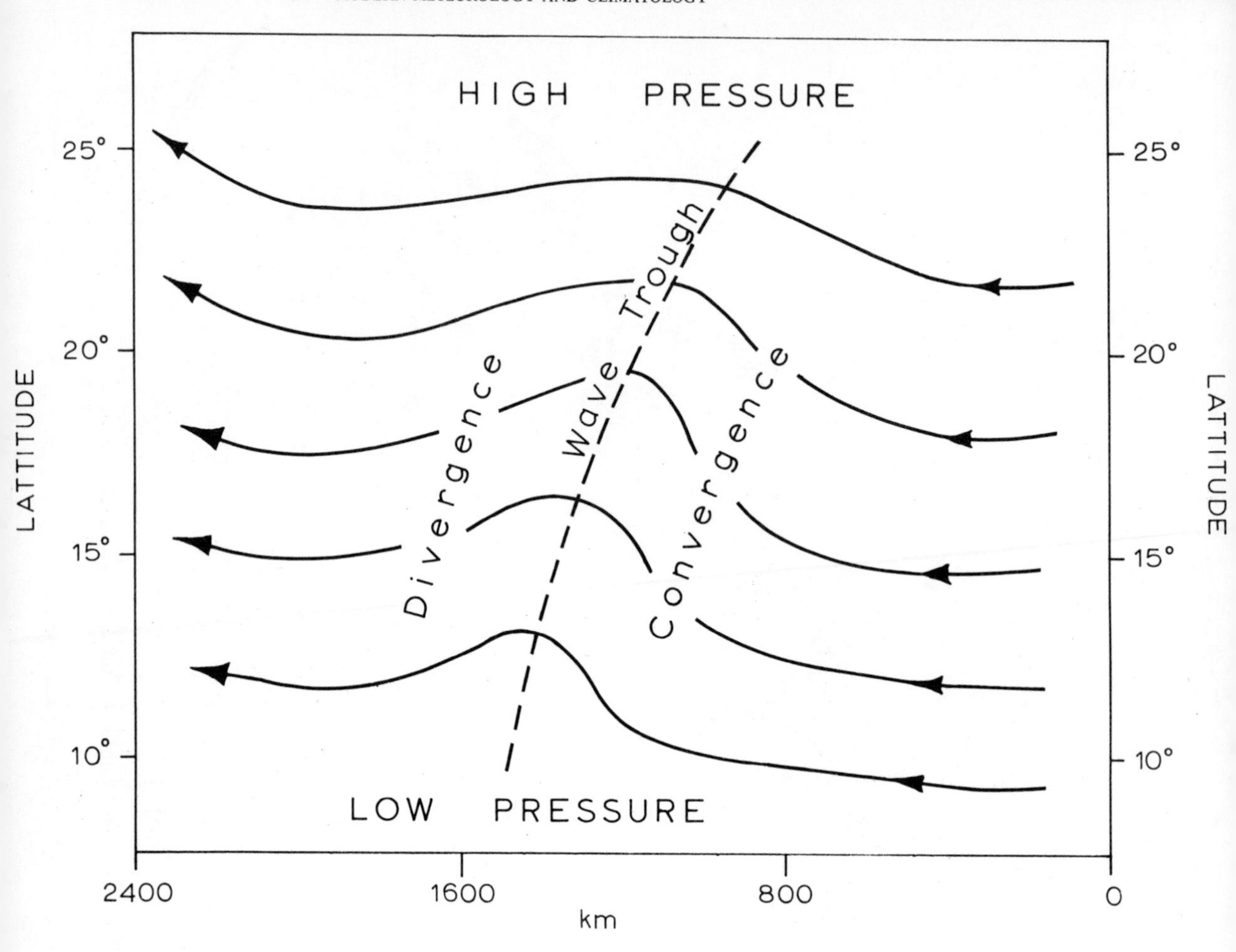

Fig. 33 Airflow at about 1500 m around a tropical easterly wave in the northern hemisphere trade winds. After diagram in R. G. Barry and R. J. Chorley, *Atmosphere, Weather and Climate* (Methuen, 1968).

The South-east Asia monsoon

The south-east Asia monsoon undoubtedly represents the major disturbance of any hemispheric model of the general circulation. The two hemispheres do not work independently and there are enormous energy transfers from the winter to the summer hemisphere in the sector between about 60° and 180°E. This is especially true during the summer monsoon period in south-east Asia.

The seasonal displacement of the main pressure and wind systems depends primarily upon fluctuations in the direct heating of the air by the earth. In the zone between 30°N and 30°S, this heating is about six times greater over the land than over the sea where most of the energy is used for evaporation. The intertropical convergence zone over Africa shifts from about 18°S in January to 18°N in July whilst over the Atlantic and Pacific Oceans it migrates between only about 0° and 10°N. Over India, the ITCZ reaches as far as 30°N in summer and then migrates to run across northern Australia in 21°S.

But the mechanisms of the monsoon are by no means as simple as this seasonal shift in the position of the ITCZ might suggest for this lower tropospheric phenomenon has to be viewed against the broad, three-dimensional circulation throughout the depth of the troposphere.

In winter, the broad westerly current at about 10 000 ft (3000 m) over southern Asia divides with jetstreams on either side of the Tibetan plateau (Fig. 34). The currents reunite off the east coast of China.

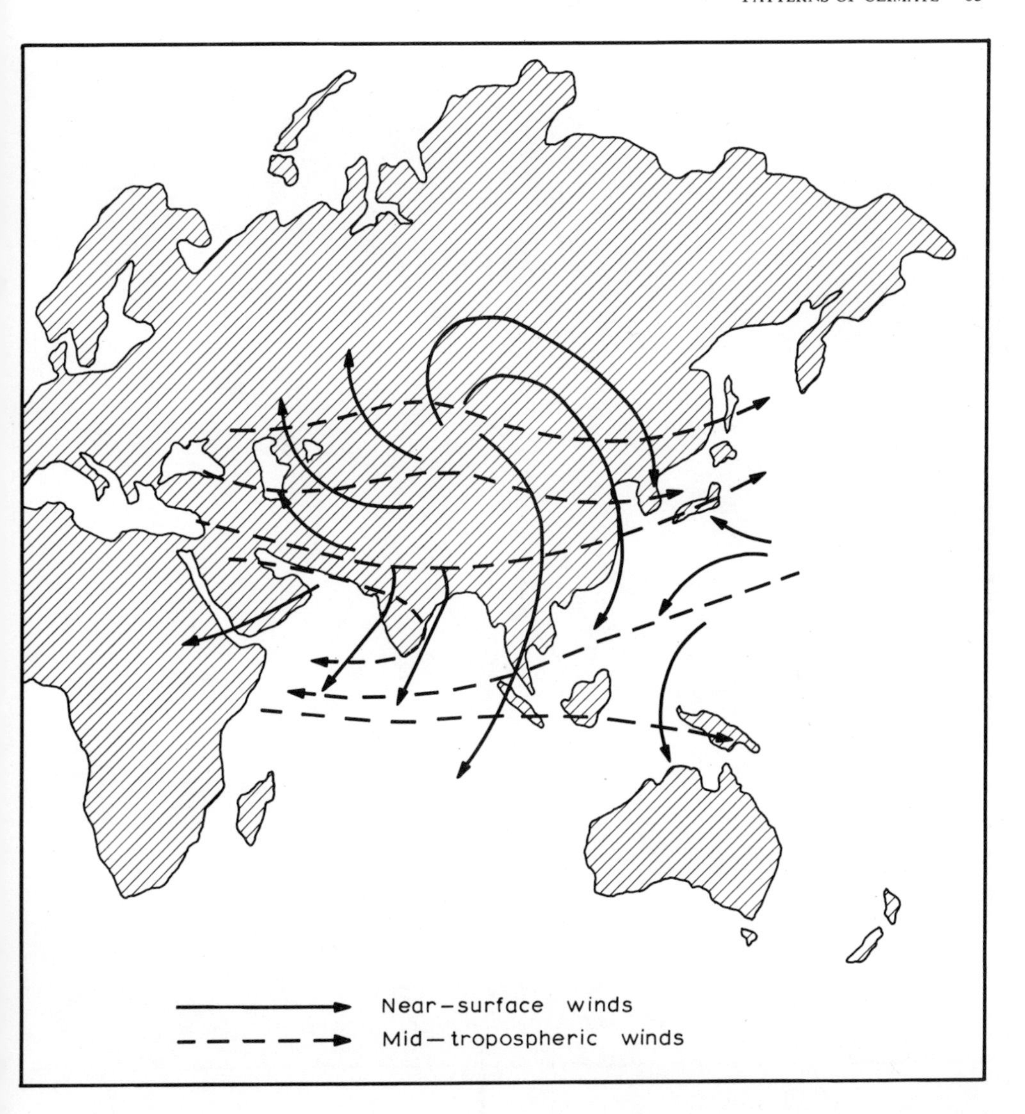

Fig. 34 Near-surface and mid-tropospheric winds over southern Asia in winter.

The southern jet is made stronger by the strong south–north thermal gradient. Air subsiding on the southern side of this jet is responsible for the subtropical anticyclone over north-west India and West Pakistan. North-westerly winds turning to north-easterly winds blow out from this anti-cyclone over India. The upper jet also steers non-frontal depressions from the Mediterranean across northern India and China.

In summer, temperatures increase above the Tibetan plateau and the westerly jet breaks down and is diverted north of the Plateau. High-level easterlies then blow over the whole of south Asia in response to the north–south thermal gradient and the formation of a high-pressure area over the Tibetan plateau at the level of its surface, about 600 mb (Fig. 35). An easterly jet also forms at about 50 000 ft (15 000 m) in about 15° latitude over southern India and the Arabian Sea. The heaviest falls of rain over India are in the area of cyclonic activity on the northern side of the jet where there is strongly ascending air. At the same time the low-level south-westerlies move northward in a series of pulses behind the equatorial trough which by mid-July reaches about 25°N. Only the lower levels of this south-westerly air-stream contain much moisture, and extensive instability and storms occur only near

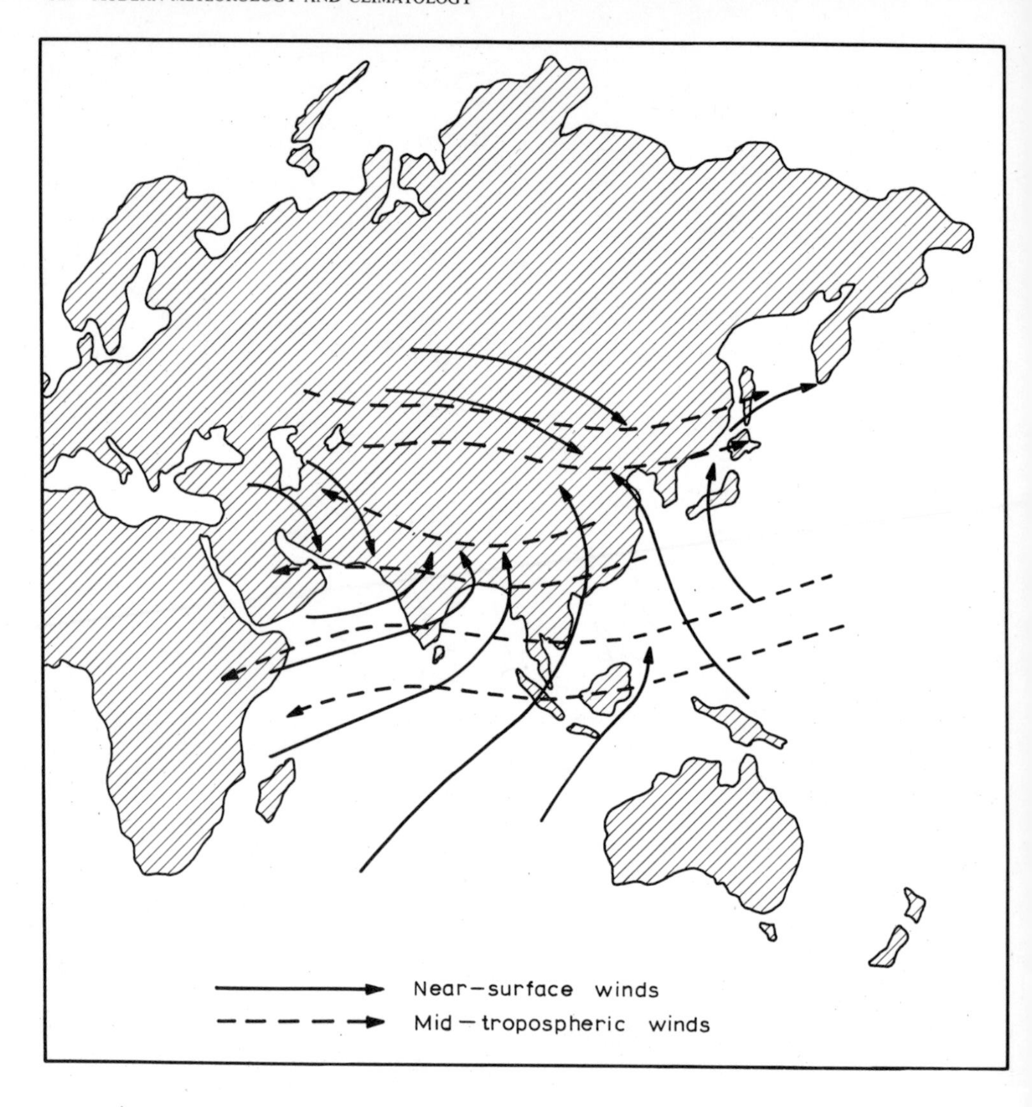

Fig. 35 Near surface and mid-tropospheric winds over southern Asia in summer.

Fig. 36 *(opposite page, top)* A section from north-west to south-east across West Pakistan and north-west India during summer showing the effect of air subsidence upon the height of clouds and thereby upon precipitation. After diagram in J. S. Sawyer, *Quart. J. R. Met. Soc.*, 1947.

the west coast of India where the airstreams converge and rise over the Western Ghats, but the moisture is soon exhausted. During the summer, the sub-tropical high lies over the Tibetan plateau giving westerly winds on its northern side. On its southern side, vertical movement in the monsoon air is very severely limited to a very shallow layer by a low inversion formed by the subsidence of continental air, and north-west and north India consequently remain dry (Fig. 36). Breaks in the monsoon occur when waves in the mid-latitude westerlies deepen in amplitude and displace the Tibetan high. Rain then falls in troughs which move eastwards over northern India but to the south, there is little rain.

Over southern China in summer, surface winds from the south-west are overlain by weak easterly winds. Much of the rain is associated with thunderstorms in shallow thermal lows.

The West African monsoon

In July and August, the ITCZ in West Africa reaches 20°N but little rain falls as far north as this because of the hot, dry Saharan easterlies which overlie the very shallow wedge of humid, cooler south-westerly winds (Fig. 37). The dry, stable Saharan air yields no precipitation during

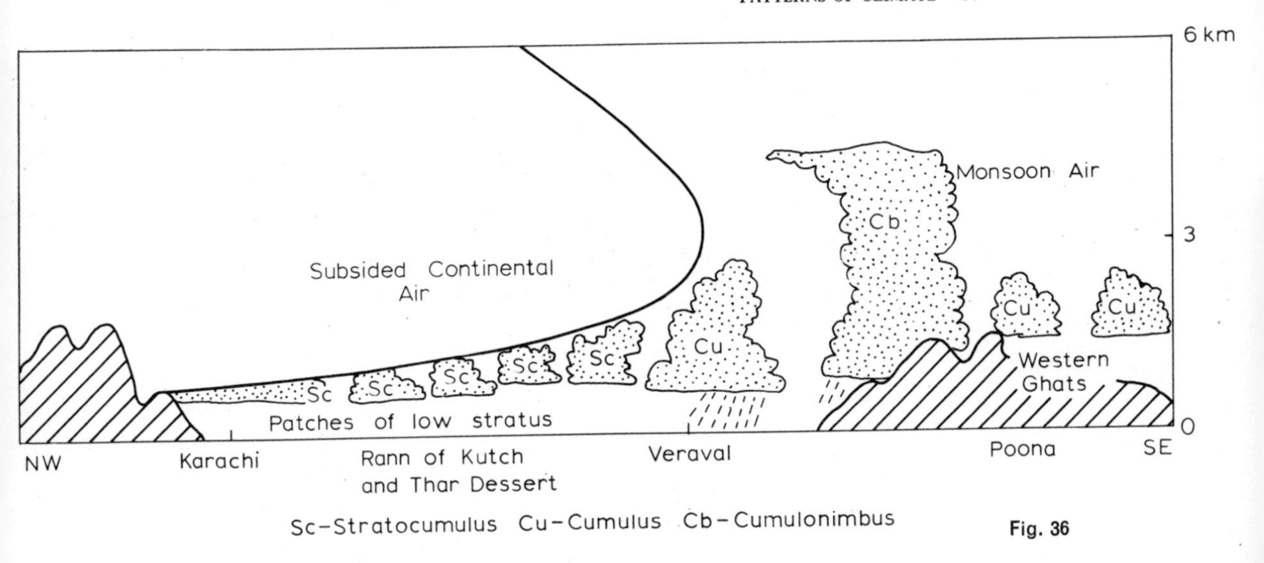

Fig. 36

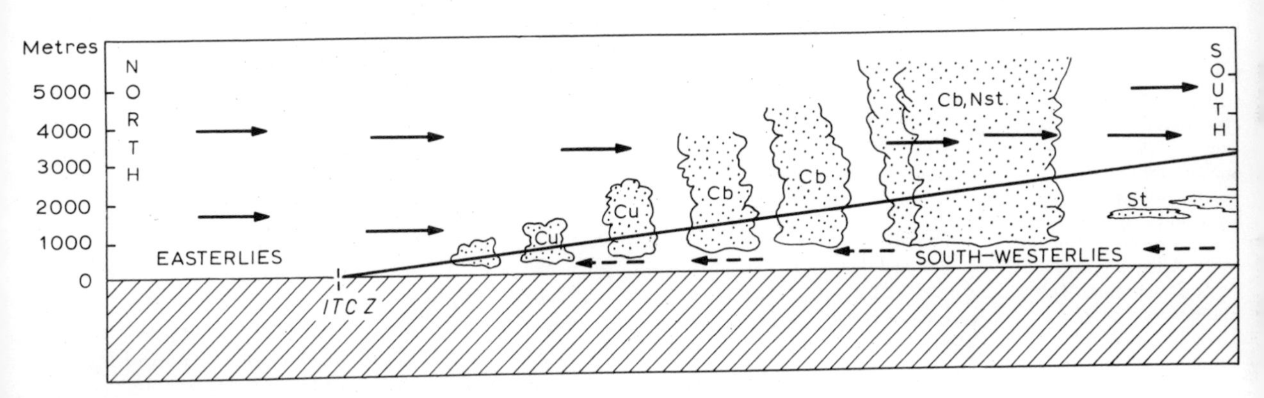

Fig. 37 North–south section through the intertropical convergence zone between the hot, dry Saharan easterlies and the warm, humid southwesterlies over West Africa in July. As the depth of the southwesterlies decreases, so rising currents and clouds are less able to penetrate the easterlies above the sloping convergence zone surface. After diagram in G. T. Trewartha, *Earth's Problem Climates* (University of Wisconsin Press, 1961).

its very shallow ascent over the ITCZ and the maritime air beneath quickly loses much of its moisture and becomes very shallow northwards from the coast. Moreover, any cumulus clouds which rise above the ITCZ are quickly dessicated in the overlying dry Saharan air. Only well south of the surface ITCZ is the maritime air sufficiently deep for the development of deep cumulonimbus clouds giving heavy rain. The surface ITCZ in West Africa in summer is thus associated with drought, not heavy rain.

North of the surface ITCZ dry, northerly winds blow in a long trajectory over the Sahara. By day they are hot, and by night they are cool but always they are dry and dusty. In winter these winds surge south behind the ITCZ. The air comes either from subsidence in an extended subtropical Azores high or from Eurasia. In the latter case it crosses the Mediterranean and underruns the warm Saharan air at a vigorous cold front which may cause low-sun period rain along the Guinea coast.

Chapter 9

Local climates

The importance of energy and moisture exchanges between the earth and the air has been noted many times. Seventy-five per cent of the atmosphere's heat input and 100 per cent of its moisture comes via or from the earth's surface and surface friction accounts for the dissipation of about 40 per cent of the atmosphere's kinetic energy. Clearly surface and near surface conditions are of paramount importance in the atmosphere's various energy budgets.

Earth-air exchanges of heat, moisture and momentum operate on a variety of areal, volumetric and time scales whose dimensions will control the intensity of the local meteorological consequences. As a result, the particular character of the earth's surface, whether or not this is in whole or part the outcome of man's activities, will often exert its strongest influence upon micrometeorological conditions of the atmosphere at times of temperature inversions in the **surface boundary layer,** being defined as the lowest 300 to 600 ft (100 to 200 m) of the **planetary boundary layer.** The latter, sometimes known as the **friction layer,** is defined as the layer extending from the earth's surface to about 2000 ft (600 m) in which the air motion is significantly affected by surface friction. The term **boundary layer meteorology** is generally applied to the study of the various energy transfers in the air-earth interface. The study of the effect of the earth's topography and surface cover upon changes of weather and climate over short distances is generally known as **local climatology** or **microclimatology** though the latter term is best restricted to the study of very small scale changes of atmospheric properties such as those within a crop.

Influence of relief on winds and temperatures

Perhaps the most fundamental local effect is that upon airflow produced by the shape of the earth's surface. Figure 38 shows the characteristic wind profiles above areas of contrasted surface friction. In all cases, mean horizontal wind speed increases with height so that winds are generally stronger in upland areas, particularly where these rise sharply above the surrounding lowlands. Air moving across extensive highland areas is also forced to concentrate its flow vertically and thereby accelerate over

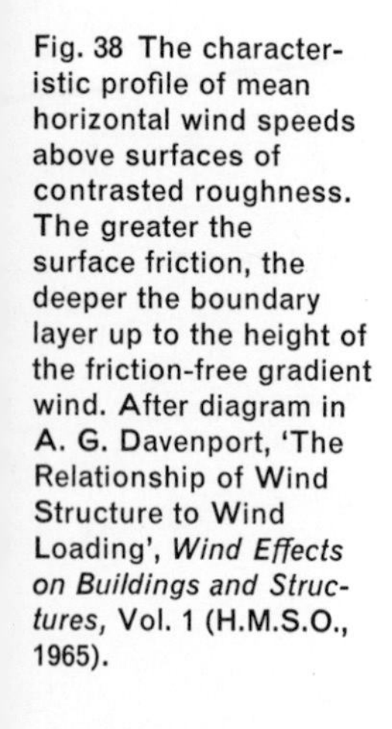

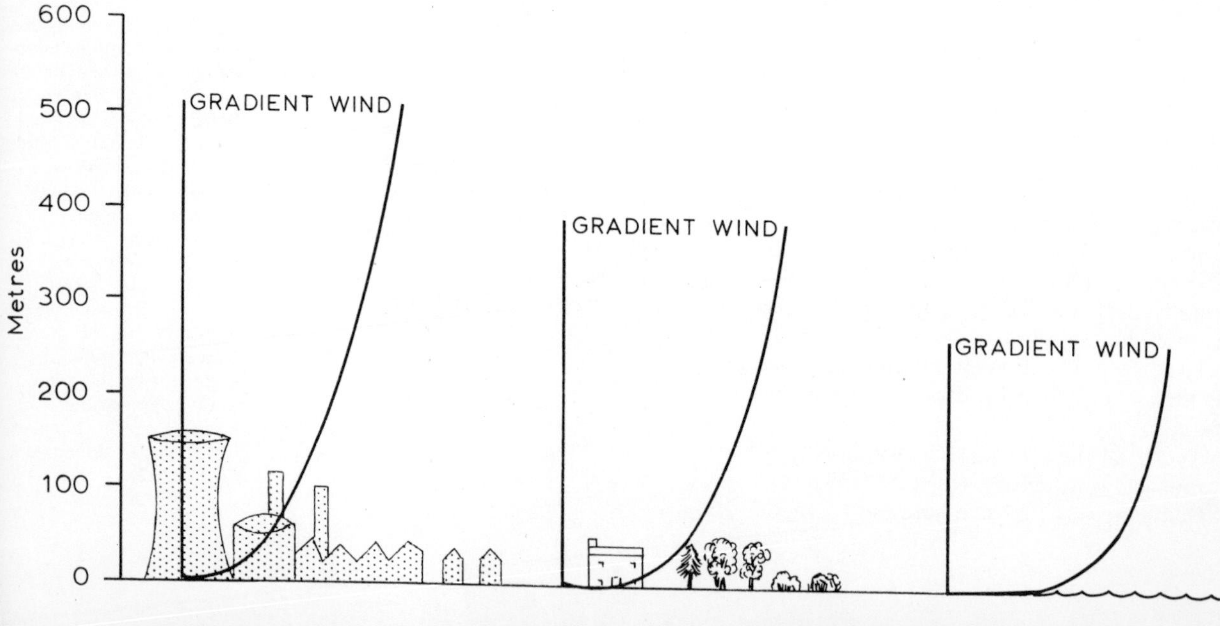

Fig. 38 The characteristic profile of mean horizontal wind speeds above surfaces of contrasted roughness. The greater the surface friction, the deeper the boundary layer up to the height of the friction-free gradient wind. After diagram in A. G. Davenport, 'The Relationship of Wind Structure to Wind Loading', *Wind Effects on Buildings and Structures*, Vol. 1 (H.M.S.O., 1965).

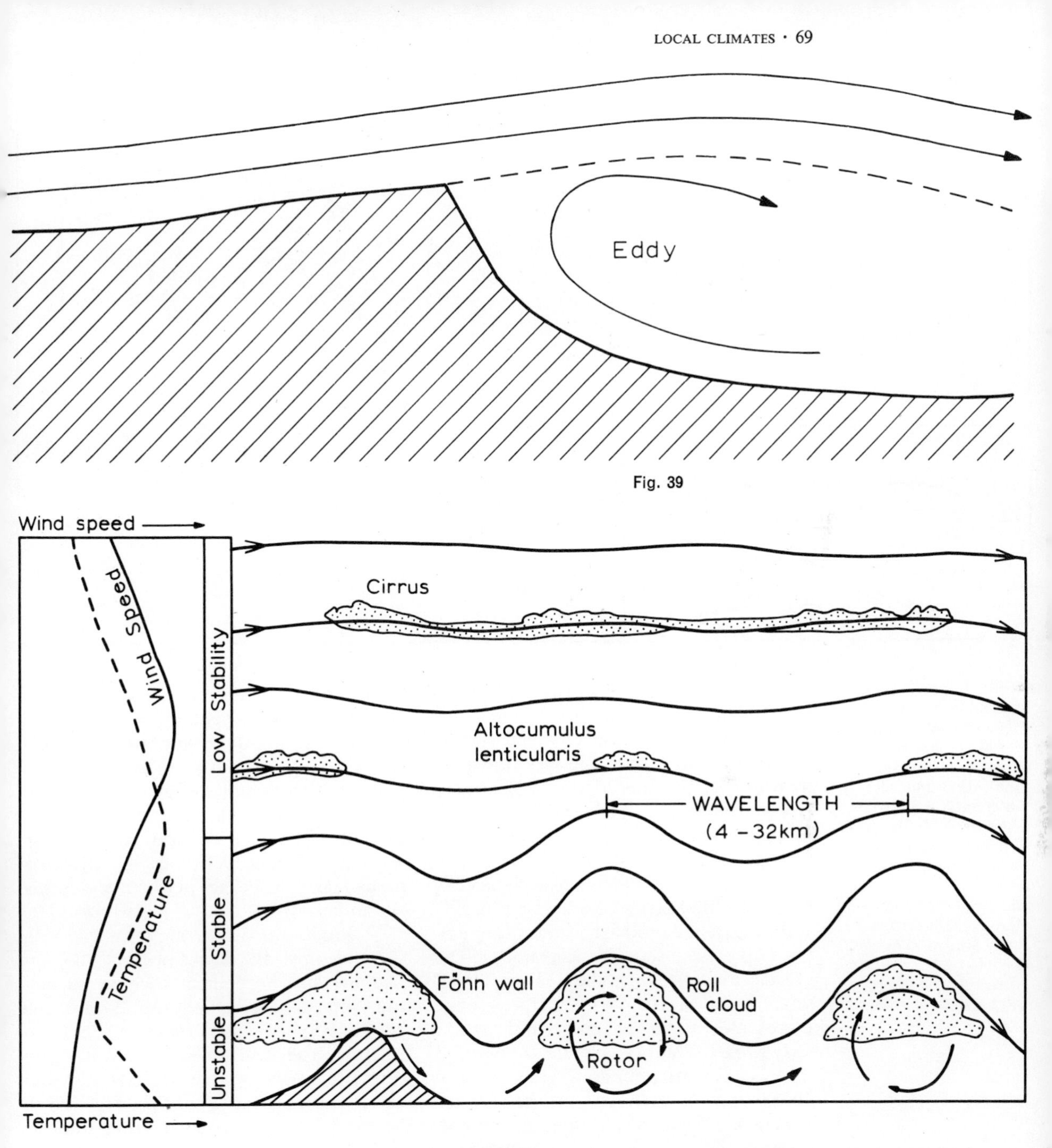

Fig. 39 *(top)* With air flowing over a hill having a sharp change of slope on the leeward side (such as a corrie), an eddy forms beneath a surface of separation, above which there is laminar flow.

Fig. 40 Lee waves and associated weather downwind of a wide topographic barrier. The strongest lee waves, having the greatest amplitude, form in the stable section of the airstream and are especially well developed when windspeeds as well as temperatures increase with height. After diagram in C. E. Wallington, *Weather*, Vol. 15, 1960.

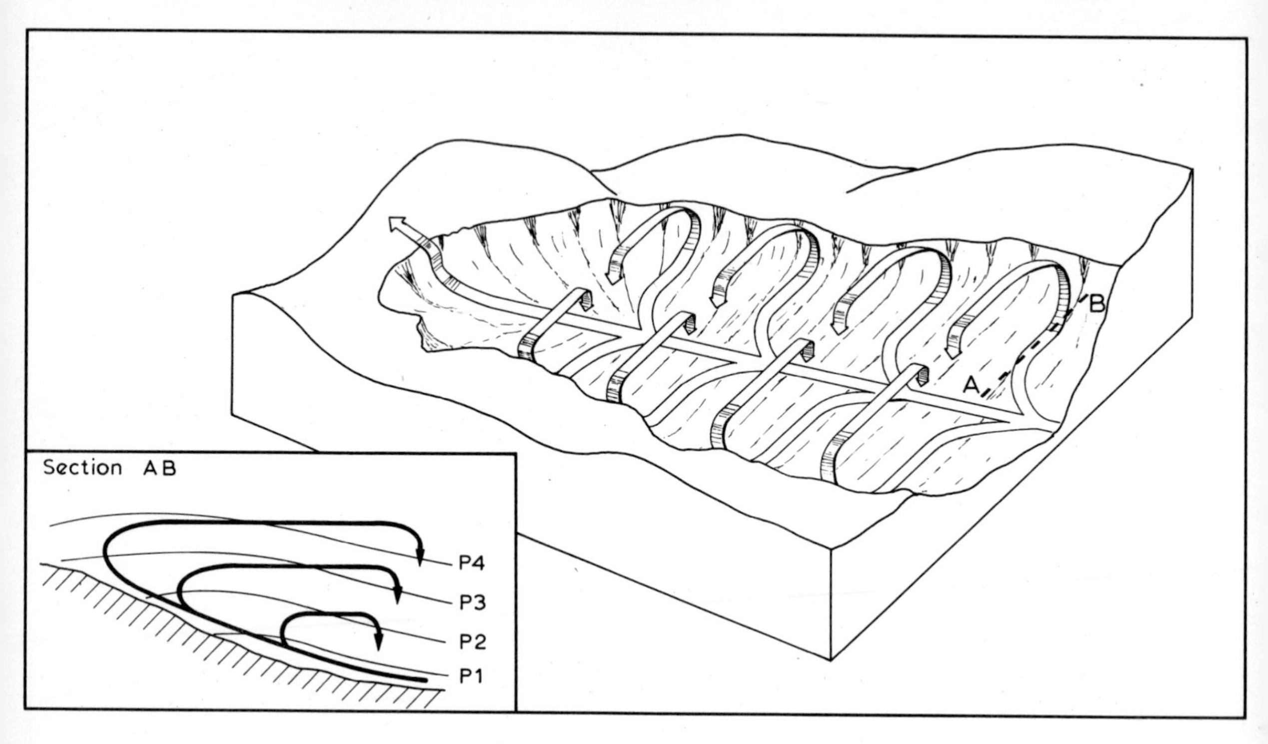

Fig. 41 Anabatic (valley) winds rising up warm valley slopes by day. The winds are maintained by the pressure gradients produced by the distortion of the pressure surfaces owing to the differences in temperature between air near the valley sides and air at the same level high above the valley. If one side of the valley is heated more than the other, the circulation will become asymmetric.

the crests. Over leeward slopes, the streamlines of the airflow diverge and the winds slacken and in any case, the near-surface winds will have been slowed down by friction over the crest. If there is a sharp change of slope, then eddies may be formed to the lee of the hill with air rising upslope before joining the regional wind at a higher level (Fig. 39). When such an eddy forms in offshore winds to the lee of a cliff, seabirds are able to soar in the rising currents up the cliff-face. If the leeward slopes are more gentle, then the airstreams will initially follow the form of the ground more closely but occasionally, and particularly when the air is stable, the topography throws the airstream into a series of waves above and downstream of the barrier. Clouds are frequently formed in the cool crests of these **lee waves** under which there are overturnings of air known as rotors (Fig. 40).

In an area of differentiated relief, the wind will be channelled and its speed accelerated through gaps and along valleys running in the direction of the wind while eddies will form across valleys lying at right angles to the wind. But mountain gap winds are often unusually warm or cold as well as strong and gusty, for mountains frequently act as barriers separating contrasted climates. The hot, dry **Santa Ana** wind which blows from the Mojave Desert through the gaps in the mountains around the Los Angeles basin, California, is one such example. The cold **Mistral** which blows down the Rhone valley to the Mediterranean coast of France in winter and spring is another.

Mountains also generate local winds in their vicinity. By day, slopes in sunlight and the air above them become warmer than air at the same level above the adjacent valley floor, plain, or slopes in shadow. The isobaric surfaces will therefore rise in the warmer, lighter, air and circulations similar to those in Fig. 41 will be established. The wind rising up the valley slope is known as an **anabatic** or **valley wind.** It is generally rather light and variable. After sunset the air over the hillside cools to temperatures below those at the same height away from the ground and the circulation is reversed (Fig. 42). The night-time downslope movements are known as **katabatic** or **mountain winds.** They are normally shallow and rarely move faster than 10 m.p.h. (16 km/h.) unless they

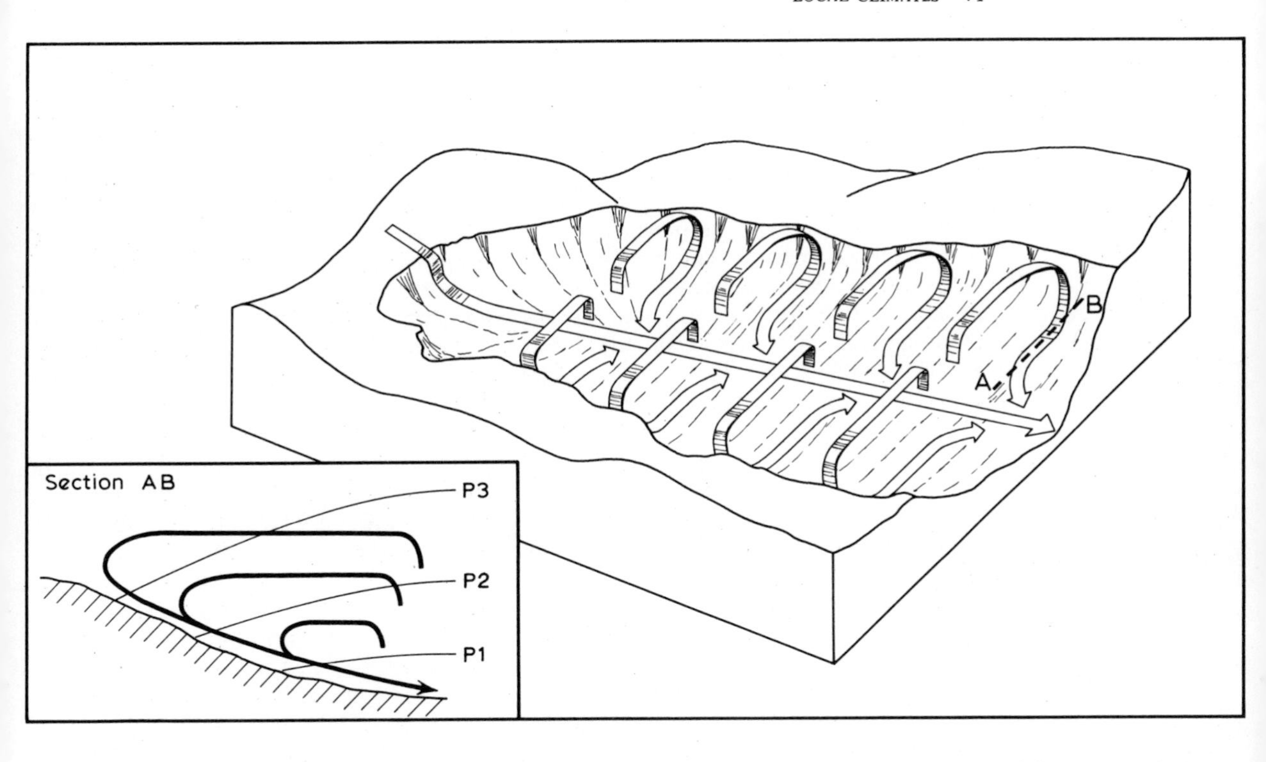

Fig. 42 Shallow, cool katabatic (mountain) winds moving downslope by night. The cool air may accumulate in valley bottoms to produce frost hollows.

are concentrated through gaps. Slopes of as little as 2° can be sufficient to cause the downward creep of cold air which then collects in the low-lying parts of the terrain to form **frost-hollows.** Cold air may also bank up behind barriers such as walls and hedges lying across its path. In areas of Britain where cold air accumulates at night, ground temperatures may fall below freezing on more than three nights in five and air temperatures on two nights in five. In deep, Alpine valleys the frequencies are even higher and night-time temperatures are frequently more than 10°C lower than those on adjacent slopes and hill-tops from which the cold air has been freely drained. By day, radiation is reflected into valley bottoms where there is limited air movement and diffusion of heat so that daytime temperatures rise sharply and there is a very large diurnal range of temperature. Air temperatures in the bottom of even modest English valleys may often range diurnally over more than 30°C which is equivalent to the difference in mean annual air temperatures between the Equator and the Arctic Circle. Ground temperatures are even more extreme.

Another thermally distinctive type of local wind is the strong, gusty, hot, dry wind which periodically descends the leeward slopes of upland areas. They are known as Föhn-type winds after the **Föhn** which blows down the northern slopes and valleys of the European Alps in winter and spring. Once the Föhn starts to blow, temperatures may rise by more than 20°C in a few minutes and the relative humidity may drop to as low as 10 per cent. Such conditions will cause rapid melting of snow and may lead to avalanches and flooding in winter and forest fires in spring. The **Chinook** wind east of the Rockies in North America, and the **Zonda** in the Andes foot-hills are similar phenomena. A slight warming can sometimes be detected in winds descending from the Scottish Highlands, the Pennines and Welsh Uplands in Britain. The explanation of the extra warmth lies party in the liberation of latent heat and loss of moisture in the airstream during the ascent of the windward slopes and partly in the warming during descent of the relatively dry, warm air of a high level inversion.

These local differences in air temperature between valley bottoms and hill crests complicate an otherwise general and fairly rapid fall of temperature with height. The rate of fall is very dependent on the synoptic meteorology of the area and the relative frequency of the various weather types, each with its own characteristic lapse rate. In Great Britain, the dominance of westerly airstreams is responsible for a remarkably sharp vertical gradient of mean temperatures. This is because these airstreams, initially very cold at all levels, are then appreciably warmed in the lower 10 000 ft (3000 m) or so by the warm waters of the north-east Atlantic. Because of this, the temperature lapse rate averages about 1°C per 500 ft (150 m) which means that it is too cold for trees to grow above about 2200 ft (670 m) (the climatic tree line) although for other reasons, such as strong winds and waterlogged shallow soils, few trees, in fact flourish above 1000 ft (305 m). In central Europe where these westerly airstreams are less common, the climatic tree line is not reached until 5000 ft (1500 m).

Rainfall also increases with height although the reasons are often more complex than is sometimes assumed. The forced uplift over upland areas not only cools the air by expansion and mixing, it also makes it more unstable. Daytime instability is also triggered off over warm, sunlit surfaces and heavy showers are therefore common in hilly areas. Cold fronts moving over upland areas are steepened and made more active. The incidence of rainfall in upland areas is consequently very variable in time and place, a complexity which is exaggerated by the time lag between cloud formation and precipitation.

Vegetation, soils and local climate

Local climates are also affected not only by the shape of the land but also by its surface covering. The nature of the vegetation cover in rural areas and of the urban fabric in towns will influence the various energy and moisture exchanges between the ground and the air. The colour of the surface will be important for this will affect the **reflectivity** or **albedo** and hence, the amount of heat absorbed. During the day dark surfaces such as coniferous and equatorial rain forests and the air above them will become warmer than lighter coloured surfaces such as grass. Beneath the crown layer of forests and within crops, shading will reduce daytime and increase night-time temperatures. Above grass however, both day and night temperatures will be lowered by evaporation and by the low thermal capacity of the air trapped between the leaves so that radiation frost is more frequent above grass than above bare soils. Winds inside forests will be very much lighter than outside and because of this, air humidities will be higher in spite of reduced amounts of precipitation reaching the ground. A lot of rain is trapped and then evaporated in the crowns of trees. Coniferous trees are especially efficient in this respect and can cause significant losses of ground water which might be serious in water catchment areas. But rainfall is not always reduced inside forests. In foggy areas trees lead to **fog drip** by causing the tiny fog droplets drifting against them to precipitate (see p. 41).

Nobody is quite sure whether or not forests are able to increase local precipitation. Some experiments show that they might but it is very difficult to be sure. One reason for our uncertainty is that it is hard to accurately measure rainfall in forests. If the effect is real, any small increase is probably explained by stronger thermals above the forest. Reported deficits of soil moisture in areas cleared of forest is more likely to result from diminished storage capacity following soil erosion rather than from any decline in precipitation caused by forest clearance.

Soil types may also be relevant to local climates, the important physical factors being albedo, thermal capacity and thermal conductivity. Dark coloured soils will tend to have higher ranges of surface temperature than light soils but clays, with their high thermal capacity per unit volume, tend to warm up and cool down less than dry sands with their large amount of trapped air and their low thermal capacity. Near the surface therefore, sands have a tendency to be warmer than clays during the day and in spring and summer. By night and in winter, however, they tend to be colder. Further down in the soil, the greater conductivity of clays gives a larger range of temperature than in sands, unless the sands are saturated, in which case

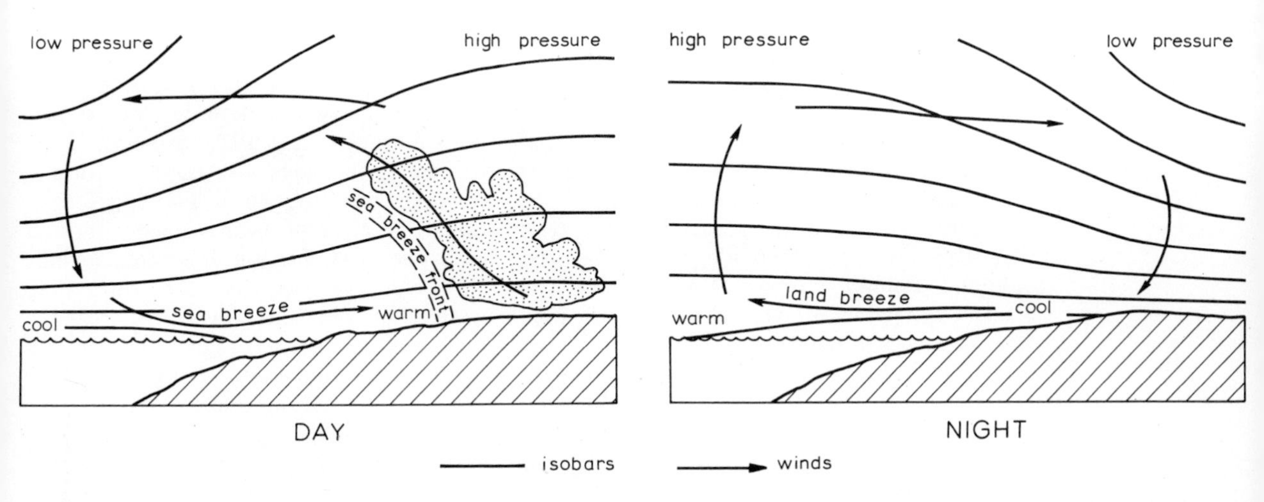

Fig. 43 By day, the air is warmer over the land than over the sea. Because of this the isobaric surfaces are distorted and a circulation across the coastline such as that shown in the left-hand diagram is produced. Near the surface the cool sea breeze moves inland in a series of pulses and is sharply separated from the warm land air by a sea breeze front. This often gives rise to clouds and showers. By night the thermal gradient across the coastline is reversed and a land breeze blows offshore near the ground.

their thermal response is more like that of clays. These characteristics are relevant to agriculture.

Seas and lakes

Lakes and seas affect local climates in a number of fundamental ways. Depending upon their size and depth, surface winds will be stronger and less turbulent owing to their relatively smooth surface; humidities will be higher and heat will be conserved and redistributed. Much will also depend upon the general atmospheric conditions, particularly the volume of air moving over and affected by the lakes and this is a function of windspeed, turbulence and stability.

Compared with conditions inland, the air above seas and coastal areas is generally cooler by day and in summer and warmer by night in winter. Because of this, coasts tend to be sunnier with less cloud than over inland areas by day and in summer. The differences in temperature can also cause daytime sea breezes and night-time land breezes as illustrated in Fig. 43. Initially the sea breeze will blow directly inland but it is generally light and unless carried by the prevailing wind, it will not be felt beyond a narrow coastal strip a few miles wide. During the course of the day, the direction of the sea-breeze will change as pressures fall more widely over the land. In the late afternoon it will be weaker and will often blow almost parallel to the coast, that is with low pressure on its left hand side (in the northern hemisphere), gradually swinging round in direction to form the night-time land breeze. A **sea-breeze front** is formed on the leading edge of the cool sea-breeze air as it pulses inland and warm land air is forced above this to give clouds and sometimes rain (Fig. 43).

Precipitation along coasts is clearly most strongly affected by their influence upon temperatures and atmospheric stability and by changes in topography such as cliffs, rather than by any differences in absolute humidity. Only the largest lakes affect mean absolute humidities and precipitation in their vicinity although on calm, clear nights the effect may be strong enough to give fogs close to quite small water bodies such as rivers and ponds.

The effect of man on local climate

Of all the various types of local climate, one of the most distinctive has been fashioned by man. This is the climate of built-up areas. By changing the thermal, hydrological and roughness properties of the surface, quite distinctive changes have been made in the atmospheric conditions between and above buildings, and in towns and cities a regionally distinct and economically important 'urban climate' is created. Once again, it is at times of light winds, clear skies and inversions in the boundary layer that the greatest modifications are made.

The three most fundamental changes are to the chemical composition of the air, to the patterns of airflow and to the temperature. A great variety of solids, liquids and gases are poured into the atmosphere from the various domestic, commercial and industrial sources that are found in most cities. But among the wealth of chemicals, three are more abundant than the rest: these are carbon dioxide, smoke particles and sulphur dioxide. The last two are measured in most countries and can be used as indices of the general levels of pollution.

The concentrations of pollution near to the ground are very uneven in both space and time, being a function of several variables. Among these variables are the quantity and density of pollution released into the atmosphere; the height, temperature and velocity at which the pollution is emitted, and the atmospheric conditions of wind and temperature which control the volume of air in which the pollution is diffused.

In many cities, the greatest quantity of smoke is emitted by domestic sources and they certainly make the major contribution to near-surface concentrations because of the low level and low temperatures and speeds at which they are discharged into the atmosphere. Smoke from domestic chimneys is brought down to the ground quite close to its source by the eddies which form around buildings. The smoke from industrial chimneys on the other hand is emitted at heights where the prevailing air currents are normally stronger and these disperse the smoke vertically and horizontally (Plate 18), often for hundreds of miles, so that mean concentrations are generally low. In contrast, large particles of grit from furnaces and similar sources very soon fall to the ground.

In many cities, emissions of smoke have been drastically reduced in recent years by the change over to gas and oil as domestic fuels. Many industries have also converted their plant to burn oil and natural gas. The increasing use of electricity also helps to clean the air by concentrating emissions at efficient plants. For these reasons and because of the requirements of clean air acts, emissions of smoke have been reduced in many countries although there have been correspondingly sharp increases in the emissions of sulphur dioxide from the burning of heavy oils by power stations and industry. But again, because of their high chimneys, near-surface concentrations have not increased proportionally.

Because the surface of built-up areas is rougher than most rural landscapes, frictional drag upon near-surface winds is increased and average horizontal windspeeds are reduced. At the same time winds between and immediately above buildings are made more gusty and variable in direction and speed so that there are fewer calms in urban areas.

Most settlements are warmer than the surrounding country. The warmth of the city, known as its **heat-island,** is generally strongest by night, although much depends upon the amount of cloud and the speed of the wind. The strongest heat-islands occur with clear skies and calm air and at these times, an inversion in the lower atmosphere will prevent the lifting of the warm air. Commonly by day but less frequently by night, the thermal lag of the city fabric gives lower temperatures in cities than in the surrounding country.

The explanation of heat-islands is not simple but most important are the differences in heat capacity and heat conductivity between city fabric and vegetation covered soils. Heat is stored in the tremendous mass of the buildings and roads, to be released more slowly than from the relatively shallow layer of soils affected by temperature changes in rural areas. The rise and fall of daily and seasonal temperatures in built-up areas has a lag behind that in rural areas and this helps to produce the temperature differences. In some cities, the direct production of heat by combustion in winter is big enough to equal the radiation received from the sun and this source of warmth must often be important in explaining the higher urban temperatures. Its efficiency is, however, reduced if, as often happens, it is released when winds are strong and deeply turbulent. In many cities of western Europe and North America, the occurrence of the strongest heat-islands in summer and early autumn indicates the limited importance of combustion in warming the city's air. In other

18. Smoke from brick-work chimneys in Bedfordshire diffusing in strong winds.

areas, in some cities in Japan for instance, combustion is thought to be the most important single factor responsible for their heat-islands. Turbulence above the city may also contribute to the higher night-time air temperatures by mixing the air within the boundary layer: in country areas, inversions with low near-surface temperatures form much more readily than in cities. Also, the very stagnant air in the bottom of streets and courtyards may also limit the loss of daytime warmth at times of moderate or strong winds.

In addition to changes in airflow, pollution and temperatures, settlements are also characterized by reduced radiation (including sunshine) and poorer visibilities because of the pollution haze; by lower relative humidities but similar or perhaps slightly higher absolute humidities; by more cloud, and sometimes by more frequent thunderstorm rains.

Chapter 10

Climatic change

As late as the nineteenth century, it was widely believed that apart from seemingly random year-to-year changes, the climates of the world had remained constant for hundreds and perhaps thousands of years. Because of this, thirty years' records were thought long enough to give representative averages of the various elements and these have been the basis of much of our agricultural, domestic and transport planning. Now we know that there are substantial fluctuations or trends of climate over a few years and decades as well as over centuries and millenia. The widespread warming during the first four decades of the present century and the cooling since about 1940 are recent evidence of the instability of our climate.

In Britain, the Quaternary Ice Age reached its peak about 17 000 years ago, but by 8000 B.C. the country was almost clear of ice. As the ice sheets melted, sea levels rose and the North Sea broke through the Straits of Dover to make Britain an island. Between 5000 and 3000 B.C. summers were generally warmer than today. After this, the climate became cooler and wetter with particularly damp and overcast summers between about 1000 and 500 B.C. Improved evidence of climatic change since then has detailed many long and short-period fluctuations. Warmer weather characterized the years between A.D. 1000 and 1250 and between 1550 and 1850 there was the so-called Little Ice Age. During this period, alpine glaciers grew and Arctic pack ice spread very far south. This was followed by another warming which lasted to about 1940 and led to the widespread melting of land ice and Arctic sea ice and a rise in sea levels. The northern hemisphere's mean air temperature, having risen by 0.6°C between 1890 and 1940, then dropped by 0.2°C by the middle 1950s. In places, and more particularly in Arctic areas, changes have been four or five times as great.

The causes of climatic change are clearly complex and multiple. No single factor can explain changes over the complete range of time scales. Some changes may be natural instabilities in the general circulation of the atmosphere, that is, it may be normal for periodicities to exist in the energy exchanges linking the lands, oceans and atmosphere and the answer to many changes of climate may be internal and not external to the earth-atmosphere system. The oceans have tremendous amounts of energy stored within them and we know all too little about how it is transported and the mechanisms by which it is exchanged with the atmosphere.

We have very little evidence of variations in the amount of insolation reaching us from the sun although our records are short and we cannot be sure that variations have not taken place in the past. Nevertheless, variations in the terms of the heat balance within the earth-atmosphere system are likely to be more important than any variation in the solar constant. Volcanic eruptions are known to reduce the amount of direct solar radiation reaching the earth although a good deal of the radiation scattered by the dust is directed forward towards the ground: the burning of vegetation and the ever-increasing combustion of fuel may be having a similar effect. Increases in the amount of carbon dioxide in the atmosphere, also caused in part by the burning of fossil fuels and the firing of vegetation has been under suspicion as yet another factor of climatic change. Carbon dioxide levels have risen by about 10 per cent in this century alone and they are still increasing. The importance of these changes arises because carbon dioxide strongly absorbs out-going terrestrial radiation in a narrow but significant infra-red waveband

and it is widely believed that this has been partly responsible for the rise in mean air temperature during the first four decades of this century. These warmings were particularly sharp in the Arctic and because of this, the meridional temperature gradient between low and high latitudes was reduced. This in turn caused fundamental changes in the general circulation system of the atmosphere with enlarged subtropical anticyclones, more frequent and persistent westerly winds in mid-latitudes and a more northerly track for mid-latitude depressions. More recently, a severe cooling in high latitudes has intensified the meridional temperature gradient and the waves in the mid-latitude westerlies have amplified, with fewer westerly winds and more frequent, cold northerly winds in Western Europe.

Summary of Britain's Climatic History During the last 15,000 years

Before 12,000 B.C.	The ice sheets over Scandinavia and northern Britain were slowly retreating with periodic halts and temporary advances. In southern Britain the climate was windy, cloudy and raw.
12,000 to 10,000 B.C.	Gradual warming with relatively dry springs and early summers.
Around 10,000 B.C.	The 'Allerød' phase when the climate became warm enough for birch trees to grow in England. Glaciers existed only in Scotland. This period ended in about 8,300 B.C. with a very sudden cooling as a result of which trees were killed and glaciers reappeared in Lake District valleys.
After 7,800 B.C.	A gradual warming in the 'Boreal' phase so that birches, pines and later elms and oaks spread north and up the hillsides.
After 6,000 B.C.	The 'Atlantic' phase. Temperatures rose to peak values between 5,000 and 3,000 B.C. when they were probably 2°C higher than today. Rainfall was also high. All the mountain glaciers finally disappeared and the sea level rose quickly as the great ice sheets melted so that by 5,000 B.C. the North Sea and Baltic had filled up and the Straits of Dover were opened and Britain became an island.
3,500 to 1,000 B.C.	The 'Sub-Boreal' phase when temperatures were a little below those in the Atlantic period but, more notably, it was drier and less windy, especially in northern Britain. Birch and pine trees again became dominant in the drier areas and there were widespread hilltop settlements.
1,000 to 500 B.C.	Rapid deterioration from which there has never been a full recovery. Cooler, cloudier and wetter than in previous period. Birch trees increased in the lowlands with oaks and alders on wetter soils. Peats replaced trees in many uplands.
Historic period	Following a small amelioration and a subsequent deterioration in Roman times, the climate generally improved and this warm period reached its peak between A.D. 1000 and 1250 when storms were less frequent than either previously or subsequently. Around 1250 there were numerous great storms and frequent North Sea floods. During the following 600 years, there was a general climatic deterioration and the period from 1550 to 1850 is sometimes known as the Little Ice Age. During this time, glaciers in Europe advanced and Arctic sea ice spread very far south. Snowbeds also became common on north facing slopes in the Scottish Highlands. After 1850, temperatures rose again until the mid 1940s, since when temperatures have fallen quite sharply with a marked decline in the frequency of westerly types of circulation.

Chapter 11

Climate control

19. Cloud seeding experiment on cumulus clouds in New South Wales, Australia. Before seeding, clouds had their base at 3300 m, top at 7000 m and freezing level at 5500 m. A) (*opposite*) Nine minutes after 100 kg. of dry ice had been dropped into one cloud, radar echoes indicated rain drops and the cloud began to rise. B) (p. 80) Thirteen minutes after seeding, the cloud top rose to 9000 m and heavy rain started to fall. C) (p. 81) Cloud eventually formed an anvil shape with top between 9000 and 12 000 m.

There is no need to labour the fundamental importance of climate to man. It is a dominant element of his environment and a powerful factor of his own well being. The crops he can grow, the houses he must build, the ways he can travel and the diseases he may contract are all subject to climatic influences. Climate is also a factor in the success of many of man's economic activities. It is not therefore surprising that one of man's oldest environment-oriented dreams has been to purposefully manage the weather. But apart from measures to improve the night-time micro-climate of plants by such devices as orchard heaters, large vaned fans (which bring down the warm air of a radiation inversion) and water sprays (by which latent heat is released into the air upon freezing), only a relatively few techniques have met with any success. Among these is the augmentation of precipitation by the introduction into clouds of freezing nuclei, salt particles or water droplets, processes known as **cloud seeding** (Plate 19). Solid carbon dioxide pellets, known as dry-ice, and silver iodide smoke can certainly induce certain clouds to grow and precipitate by triggering off the freezing stage of the Bergeron-Findeisen process (see p. 41). But the techniques are clearly more complicated, less productive and less certain of success than was thought in the early years of scientific rainmaking following World War II. Nevertheless, modest increases of precipitation of between 10 and 20 per cent have been widely reported. Equally however, several experiments have resulted in a widespread decrease in rain as the result of attempts to increase it. The amount of seeding material used and the prevailing meteorological conditions are clearly very critical to success.

Cloud seeding is also being used in experiments in lightning suppression and to see if it can change the structure and movement of hurricanes. The experiments are incomplete and nobody is really sure if we can modify such a large and powerful meteorological system.

The suppression or mitigation of hail damage by the use of rockets and shells filled with silver iodide is a highly organized system of weather control in a number of countries: more than 50 000 rockets are fired annually into storms in northern Italy for instance. If the system works, it probably does so by the diffusion of ice nuclei to freeze the supercooled droplets essential for hailstone growth and by the use of the explosion to shatter hailstones along internal planes of natural weakness.

Fog clearance is another avenue of research and limited practice: many techniques have been tried but only a few have succeeded (Plate 20). The most effective method uses heat such as that from a jet engine, to raise the temperature of the foggy air along runways above the dew point. Supercooled fogs can be cleared by spraying them with silver iodide particles whereupon some of the droplets change into ice crystals; they then grow at the expense of the water droplets and fall to the ground.

On the scale of true climatic control, one must remember the enormity of natural energies compared with man's capabilities, so that frequently the only possibilities are to find means whereby relatively small energy inputs can be used to trigger off natural instabilities in a desired manner. Many spectacular schemes have been proposed. These include spreading sheets of black plastic over arctic pack ice to reduce the albedo and hence raise local temperatures; altering the courses of certain ocean currents by, for instance, damming the Bering Straits, and the creation of vast inland seas in Siberia and central Africa.

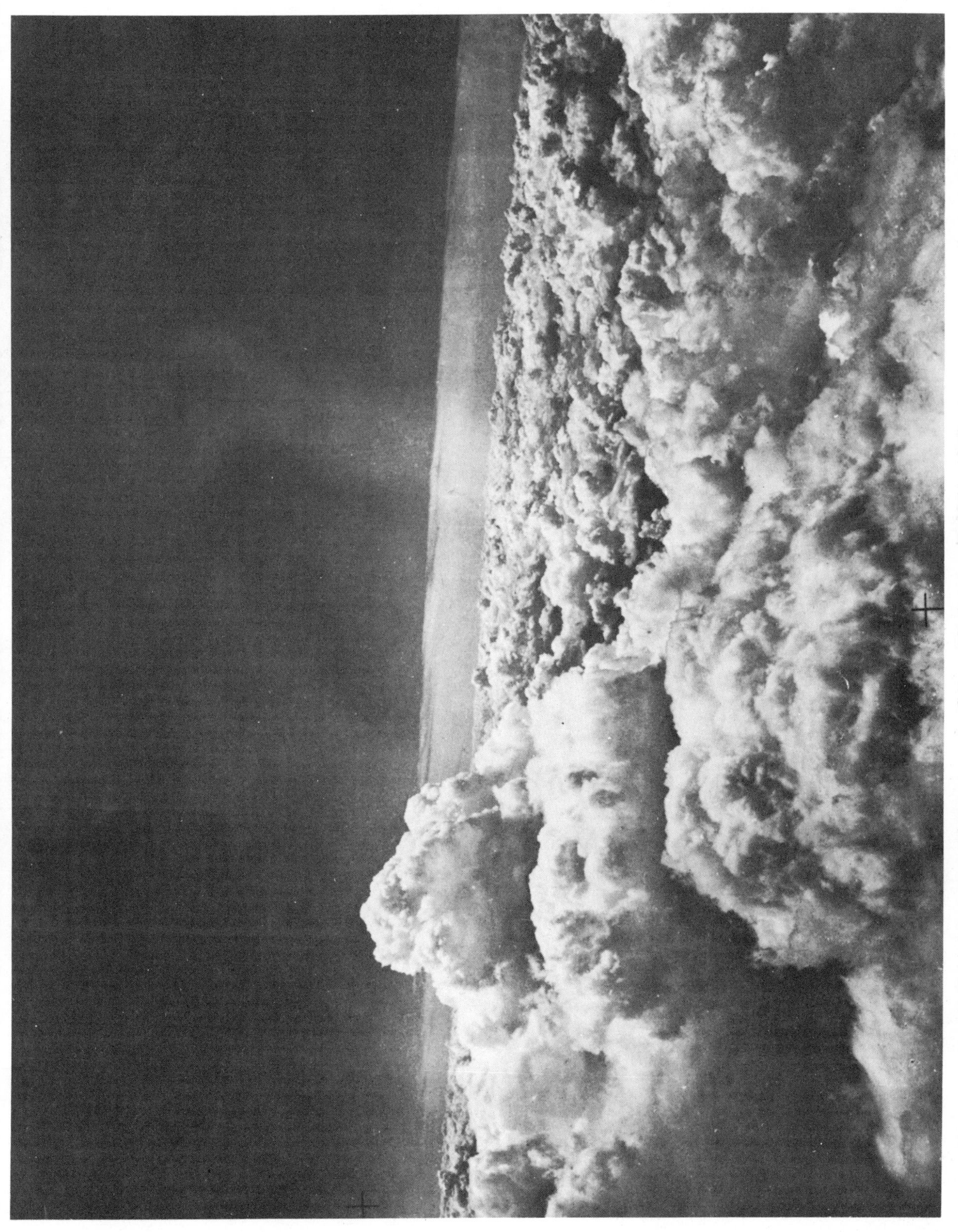

Plate 19A

Plate 19B

Plate 19C

These are wild, speculative schemes, more often than not as uncertain in consequences as in method, Similarly, large schemes affecting the physical properties of extensive areas of the earth's surface might have climate consequences which may be overlooked or unforeseen. The recent Russian scheme to divert part of the flow of Siberian rivers draining into the Arctic Ocean is a case in point. The water will be used to irrigate the lands of south-central USSR, where the over-use of ground water has seriously lowered the water table; but in so doing, the Arctic Ocean will lose part of one of its major sources of fresh water. This normally lies no more than 600 ft (about 200 m) deep above the more saline deep and bottom waters of the Arctic Ocean and plays a major role in controlling the extent of sea ice and thereby of air temperatures in the Arctic. If the supply of fresh water were seriously diminished, some of the pack ice would melt and the Arctic would become warmer. The process might then become self-developing and there is a very real chance that the general circulation of the whole atmosphere could be disturbed, the consequences of which might very well be to reduce precipitation in the very region the scheme was originally intended to irrigate.

In most cases we are not in a position to know what the full consequences of such major changes in the earth's geography would be and it would be foolhardy to find out by field experiment. Before we can so rashly tamper with climate, we must understand the workings of the atmosphere with a great deal more precision than we do now. Before we can safely experiment with the atmosphere, we must be able to predict the consequences with a high degree of accuracy and before we can do this we must better understand the exact science of meteorology.

20. An R.A.F. Lancaster bomber taking off in heavy fog from Manston, Kent in 1944. Burners were used to clear the fog and the system was known as FIDO (Fog Investigation and Dispersal Operation).

Bibliography

Barrett, E. C. *Viewing Weather from Space* (Longmans, London, 1967)

Barry, R. G. and Chorley, R. J. *Atmosphere, Weather and Climate* (Methuen, London, 1971)

Chandler, T. J. *The Climate of London* (Hutchinson, London, 1965)

Crowe, P. R. *Concepts in Climatology* (Longman, London, 1971)

Dobson, G. M. B. *Exploring the Atmosphere* (Oxford University Press, 1968)

Geiger, R. *The Climate Near the Ground* (Harvard University Press, Cambridge, Mass., 1965)

Hare, F. K. *The Restless Atmosphere* (Hutchinson, London, 1961)

Lamb, H. H. *The English Climate* (English Universities Press, London, 1964)

Landsberg, H. E. *Physical Climatology* (Gray Printing Co., Pennsylvania, 1964)

Landsberg, H. E. (Ed.) *World Survey in Climatology*, 15 volumes, various dates (Elsevier, London)

Manley, G. *Climate and the British Scene* (Collins, London, 1952)

Mason, B. J. *Clouds, Rain and Rainmaking* (Cambridge University Press, 1962)

Maunder, W. J. *The Value of the Weather* (Methuen, London, 1970)

Pedgley, D. E. *A Course in Elementary Meteorology* (H.M.S.O., London, 1962)

Riehl, H. *Introduction to the Atmosphere* (McGraw-Hill, New York, 1965)

Scorer, R. S. and Wexler, H. *Cloud Studies in Colour* (Pergamon Press, London, 1967)

Strahler, A. N. *Introduction to Physical Geography* (Wiley, New York, 1965)

Sutcliffe, R. C. *Weather and Climate* (Weidenfeld and Nicolson, London, 1966)

Sutton, O. G. *Understanding the Weather* (Pelican, Harmondsworth, 1960)

Taylor, J. A. and Yates, R. A. *British Weather in Maps* (Macmillan, London, 1967)

Trewartha, G. T. *The Earth's Problem Climates* (McGraw-Hill, New York, 1961)

Index